COMPENDIUM OF
MACROMOLECULAR NOMENCLATURE

International Union of Pure and Applied Chemistry
Macromolecular Division
Commission on Macromolecular Nomenclature

Compendium of Macromolecular Nomenclature

Prepared for publication by
W.V. Metanomski of the
Chemical Abstracts Service
Columbus, Ohio, USA

Oxford
Blackwell Scientific Publications
London Edinburgh Boston Melbourne Paris Berlin Vienna

and published for them by
Blackwell Scientific Publications
Editorial offices:
Osney Mead, Oxford OX2 0EL
25 John Street, London WC1N 2BL
23 Ainslie Place, Edinburgh EH3 6AJ
3 Cambridge Center, Cambridge
Massachusetts 02142, USA
54 University Street, Carlton
Victoria 3053, Australia

Other editorial offices:

Arnette SA
2, rue Casimir-Delavigne
75006 Paris
France

Blackwell Wissenschaft
Meinekestrasse 4
D-1000 Berlin 15
Germany

Blackwell MZV
Feldgasse 13
A-1238 Wien
Austria

First published 1991

The origination of the chemical structures
and the typesetting were done at
Interprint Limited, Malta
Printed in Great Britain
by The Alden Press, Oxford
Bound by Green Street Bindery, Oxford

British Library
Cataloguing in Publication Data

Compendium of macromolecular nomenclature.
1. Polymers. Terminology
I. International Union of Pure and Applied Chemistry
(Commission on Macromolecular Nomenclature)
II. Metanomski, W.V.
547.70014

ISBN 0-632-02846-7
0-632-02847-5 (pbk)

DISTRIBUTORS

Marston Book Services Ltd
PO Box 87
Oxford OX2 0DT
(*Orders*: Tel: 0865 791155
Fax: 0865 791927
Telex: 837515)

USA
Blackwell Scientific Publications, Inc.
3 Cambridge Center
Cambridge, MA 02142
(*Orders*: Tel: (800) 759-6102)

Canada
Oxford University Press
70 Wynford Drive
Don Mills
Ontario M3C 1J9
(*Orders*: Tel: (416) 441-2941)

Australia
Blackwell Scientific Publications
(Australia) Pty Ltd
54 University Street
Carlton, Victoria 3053
(*Orders*: Tel: (03) 347-0300)

Library of Congress
Cataloging-in-Publication Data

Metanomski, W.V.
Compendium of macromolecular nomenclature/prepared for publication by W.V. Metanomski.
p. cm.
At head of title: International Union of Pure and Applied Chemistry, Macromolecular Division, Commission on Macromolecular Nomenclature.
Includes bibliographical references.
ISBN 0-632-02846-7.—ISBN 0-632-02847-5 (pbk.)
1. Macromolecules—Nomenclature. 2. Polymers—Nomenclature.
3. Macromolecules—Terminology. 4. Polymers—Terminology.
I. International Union of Pure and Applied Chemistry. Commission on Macromolecular Nomenclature.
II. Title.
QD380.6.M37 1991
547.7'014–dc20

Contents

Preface

In the late 1960s, the International Union of Pure and Applied Chemistry (IUPAC) formally established the Commission on Macromolecular Nomenclature (Commission IV.1), which has since become the leading nomenclature body in the polymer field. The Commission promulgates rules and definitions that place polymer terminology and nomenclature on a much more systematic basis than had previously been the case (see refs 1–13 on pp. 9–10 for a list of IUPAC nomenclature reports, including several from the Commission's predecessor). The International Organization for Standardization (ISO), primarily through its Technical Committee TC/61-Plastics, the Chemical Abstracts Service and various national nomenclature bodies (such as that of the American Chemical Society) have also helped to shape the field.

In recent years, the IUPAC Commission on Macromolecular Nomenclature has developed sets of definitions for many of the basic terms dealing with polymer molecules, assemblies of polymer molecules, polymer solutions and polymer crystals. It continues to work to develop definitions of terms relating to polymer melts and solids and polymerization reactions. The Commission is also extending existing nomenclature to more complicated cases, such as irregular polymers, cross-linked polymers, double-strand polymers, interpenetrating networks and multicomponent polymer systems.

The publication of this Compendium marks the completion of the first phase of this work. There now exists a significant body of definitions and systematic polymer nomenclature. Following the IUPAC practice in other fields of chemistry, existing rules are collected here for the convenience of the scientific community.

Since this is a first major compilation of the Commission, we have included a brief history of the Commission, as well as a list of members serving on the Commission since its inception in 1968 under the Chairmanship of Kurt L. Loening.

An introduction to macromolecular nomenclature highlights the important precepts and describes them in a fashion less formal than that customarily employed in the official rules and recommendations. The documents reprinted in full in the Compendium are grouped under two major headings: definitions and nomenclature. A bibliography of biopolymer-related IUPAC-IUB nomenclature recommendations is provided in the Appendix.

IUPAC nomenclature and terminology compendia in other

fields of chemistry have acquired distinct trivial designations, derived from the colour of the cover: 'Blue Book'—organic chemistry, 'Red Book'—inorganic chemistry, 'Orange Book'—analytical chemistry, 'Green Book'—physical chemistry, and 'Gold Book'— overall chemistry.

A purple colour has been chosen for the cover of the macromolecular chemistry compendium. We hope that this 'Purple Book', as well as its future editions, will contribute to a clearer communication on the structure, reactions and properties of macromolecules. We invite comments, criticism and suggestions.

AUBREY D. JENKINS *Chairman, 1977–1985*
PAVEL KRATOCHVÍL *Chairman, 1985–1991*
IUPAC Commission on Macromolecular Nomenclature

History of the Commission on Macromolecular Nomenclature

The first publication of IUPAC in the area of macromolecular nomenclature, in 1952, was by a predecessor of the Commission, namely the Subcommission on Nomenclature of the IUPAC Commission on Macromolecules [1]*. That report was a landmark in that it systematized, for the first time, the naming of macromolecules and certain symbols and terms commonly used in polymer science. Several points are notable about this report: (1) it introduced the use of parentheses in source-based nomenclature when the name of the monomer from which the polymer is derived consists of more than one word—a practice that is now widely followed; (2) it recommended an entirely new way of naming polymers according to their structure, using the suffix 'amer'–a recommendation that seems to have been almost wholly ignored; (3) last, but certainly not least, the Commission drew on the talents of such remarkable individuals as J.J. Hermans, M.L. Huggins, O. Kratky and H.F. Mark.

Ten years later, the Subcommission issued its second report [2], which dealt with the then-burgeoning field of stereoregular polymers. This important document was written by M.L. Huggins, Nobel Laureate G. Natta, V. Desreux and H.F. Mark, with the help of F. Danusso, M. Farina and M. Peraldo. Six years later, a revision of definitions in the original report appeared [3].

The next phase of the Commission, now renamed with its present title, began in 1968. Under the Chairmanship of K.L. Loening, with first L.C. Cross and later R.B. Fox as Secretary, the expanded Commission produced a series of major documents [4b, 5c, 6b] that have shaped modern nomenclature. Especially noteworthy are the 'Basic Definitions of Terms Relating to Polymers' and the key 'Nomenclature of Regular Single-Strand Organic Polymers'. The latter, originally developed by a committee of the American Chemical Society and refined by the Commission, revolutionized polymer nomenclature by providing a systematic, consistent scheme that was particularly well adapted to indexing and became the standard for *Chemical Abstracts* and major polymer journals throughout the world.

As the 1970s came to a close, A.D. Jenkins assumed the Chairmanship, with R.B. Fox continuing as Secretary to 1979, to be

*References are listed on pages 9–10.

succeeded by N.M. Bikales who served as Secretary until 1987. There followed in rapid succession a complete revision of the stereochemical definitions [7b], the terminology for molar masses in polymer science [9], the extension of structure-based nomenclature to inorganic and coordination polymers [8b], the systematization of nomenclature of copolymers [10], two key documents dealing with physicochemical terminology in the polymer field [11, 12] and a new method of classifying polymers [13]. These documents were completed under the Chairmanship of P. Kratochvíl, who assumed these duties in 1985. W.J. Work was elected Secretary in 1987.

Throughout its existence, the Commission was fortunate in being able to attract outstanding polymer scientists from many countries. Not content with merely refining existing practice, the Commission is now engaged in pioneering work dealing with the nomenclature of irregular, double-strand (ladder and spiro) and crosslinked polymers, as well as that of polymer blends and interpenetrating networks. It also works on graphic representations for polymer structures and on the terminology of bulk polymers, liquid crystals and polymerization reactions.

NORBERT M. BIKALES *Secretary, 1979–1987*
IUPAC Commission on Macromolecular Nomenclature

Membership of the Commission on Macromolecular Nomenclature

Titular Members

G. Allegra (Italy)	1977–1989
R.E. Bareiss (FRG)	1983–1991
N.M. Bikales (USA)	1977–1987
	(*Secretary:* 1979–1987)
P. Corradini (Italy)	1968–1977
L.C. Cross (UK)	1968–1975
	(*Secretary:* 1968–1973)
R.B. Fox (USA)	1968–1979
	(*Secretary:* 1973–1979)
K. Hatada (Japan)	1989–1993
A.D. Jenkins (UK)	1975–1985
	(*Chairman:* 1977–1985)
V.V. Korshak (USSR)	1968–1971
P. Kratochvíl (Czechoslovakia)	1979–1991
	(*Chairman*: 1985–1991)
K.L. Loening (USA)	1968–1977
	(*Chairman:* 1968–1977)
I. Mita (Japan)	1979–1989
C. Noël (France)	1985–1993
I.M. Papisov (USSR)	1979–1987
N.A. Platé (USSR)	1971–1979
W. Ring (FRG)	1971–1981
V.P. Shibaev (USSR)	1987–1991
P. Sigwalt (France)	1975–1983
G.J. Smets (Belgium)	1968–1975
R.F.T. Stepto (UK)	1989–1993
C. Suhr (FRG)	1968–1971
U.W. Suter (Switzerland)	1981–1991
T. Tsuruta (Japan)	1968–1979
W.J. Work (USA)	1987–1991
	(*Secretary:* 1987–1991)

Associate Members

J.V. Alemán (Spain)	1987–1991
N.M. Bikales (USA)	1975–1977
D. Braun (FRG)	1981–1989
J.M.G. Cowie (UK)	1983–1987
L.C. Cross (UK)	1975–1979

L.G. Donaruma (USA)	1977–1985
R.B. Fox (USA)	1979–1983
K. Hatada (Japan)	1987–1989
A.D. Jenkins (UK)	1985–1987
J. Kahovec (Czechoslovakia)	1987–1991
P. Kratochvíl (Czechoslovakia)	1977–1979
K.L. Loening (USA)	1977–1985
R.H. Marchessault (Canada)	1977–1981, 1983–1987
W.V. Metanomski (USA)	1987–1991
I. Mita (Japan)	1977–1979, 1989–1991
I.M. Papisov (USSR)	1987–1991
S. Penczek (Poland)	1979–1983
N.A. Platé (USSR)	1979–1983
R.P. Quirk (USA)	1983–1987
M. Rinaudo (France)	1981–1985
T. Saegusa (Japan)	1989–1993
L. Shi (PRChina)	1987–1991
P. Sigwalt (France)	1983–1987
R.F.T. Stepto (UK)	1987–1989
L.-O. Sundelöf (Sweden)	1981–1985
U.W. Suter (Switzerland)	1979–1981
S.K. Tripathy (USA)	1987–1991
T. Tsuruta (Japan)	1979–1983
R.E. Wetton (UK)	1977–1981
H. Wilski (FRG)	1975–1979
W.J. Work (USA)	1985–1987

National Representatives

T.-O. Ahn (S. Korea)	1989–1991
D. Balköse (Turkey)	1987–1991
M. Barón (Argentina)	1987–1991
D.D. Deshpande (India)	1985–1987
Y. Hu (PRChina)	1981–1983
K. Kishore (India)	1989–1991
G.J. Leary (New Zealand)	1981–1987
K.L. Loening (USA)	1985–1987
E.M. Macchi (Argentina)	1983–1987
E.B. Mano (Brazil)	1979–1985
N.A. Platé (USSR)	1983–1991
L. Shi (PRChina)	1983–1987
A.S. Tan (Malaysia)	1983–1987
F. Tüdős (Hungary)	1985–1987
M. Van Beylen (Belgium)	1985–1987

Introduction to Macromolecular Nomenclature

PREAMBLE

This chapter is designed to offer the reader a general introduction to the principles of polymer nomenclature and to serve as a guide to the more detailed chapters that follow. Nomenclature, as used here, refers to the naming of polymeric materials. The nomenclature of scientific communication is emphasized, although there is generally little reason for differences between scientific and other, e.g. commercial usage. The nomenclature of biological macromolecules, which is highly specialized, is not discussed here. The reader is referred to the bibliography of biopolymer-related nomenclature recommendations listed in the Appendix.

BASIC DEFINITIONS

No nomenclature document is more fundamental to a given science than the definitions of basic terms used in that area. The IUPAC Commission on Macromolecular Nomenclature published a document in 1974 [4b] that offers definitions of 53 terms, including polymer, constitutional unit, monomer, polymerization, regular polymer, tactic polymer, block polymer, graft polymer, monomeric unit, degree of polymerization, addition polymerization, condensation polymerization, homopolymer, copolymer, bipolymer, terpolymer, copolymerization and many others. Both structure-based and process-based definitions are given. These definitions are essential to any systematic polymer nomenclature and the reader is strongly encouraged to consult the full document.

SOURCE-BASED NOMENCLATURE

Traditionally, polymers have been named by attaching the prefix 'poly' to the name of the real or assumed monomer (the 'source') from which it is derived. Thus, polystyrene is the polymer made from styrene, and it will often be found in an index under 'styrene, polymer of'. When the name of the monomer consists of two or more words, parentheses should be used [5d], as in poly(vinyl acetate), poly(methyl methacrylate), poly(sodium styrenesulfonate), etc. Failure to use parentheses can lead to ambiguity: polychlorostyrene can be the name of either a polychlorinated (monomeric) styrene or of a polymer derived from chlorostyrene; polyethylene oxide can refer to either the polymer (1), the polymer (2) or the macrocycle (3).

This chapter is based on the article 'Nomenclature', by N.M. Bikales, which appeared in Vol. **10** of the *Encyclopedia of Polymer Science and Engineering*, 2nd Edn, John Wiley & Sons, New York, 1987, pp. 191–204. Permission of the publisher is gratefully acknowledged.

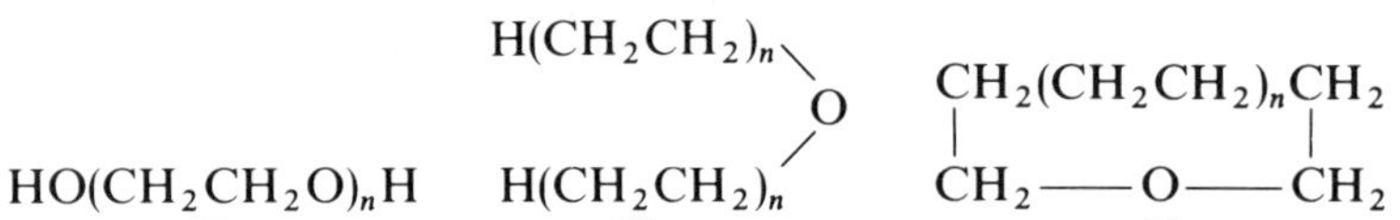

(1) (2) (3)

These problems are easily overcome with parentheses: names such as poly(chloro)styrene, poly(chlorostyrene), and poly(ethylene oxide) clearly indicate the part of the name to which the prefix 'poly' refers. The omission of parentheses is, unfortunately, still seen in the literature.

The principal deficiency of source-based nomenclature is that the chemical structure of the monomeric unit in a polymer is not identical to that of the monomer (e.g. $—CH_2—CHX—$ versus $CH_2{=}CHX$); thus, the name 'polymonomer' is actually a misnomer. The structure of the constitutional repeating unit is also not specified in this scheme; for example, polyacrolein does not indicate whether the vinyl or the aldehyde group has polymerized.

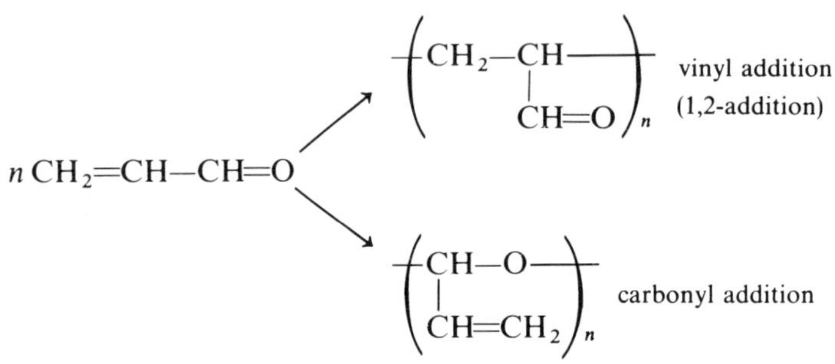

Different types of polymerization can take place with many other monomers, depending on the polymerization conditions. Furthermore, a name such as poly(vinyl alcohol) refers to a hypothetical source, since this polymer is obtained by hydrolysis of poly(vinyl acetate).

In spite of these serious deficiencies, source-based nomenclature is still firmly entrenched in industrial literature and, to a lesser extent, in scientific communication. It originated at a time when polymer science was less developed and the structure of most polymers was ill defined. The rapid advances now being made in the structure determination of polymers will gradually shift the emphasis of polymer nomenclature away from starting materials, toward the structure of the macromolecules.

Copolymers

Copolymers are polymers that are derived from more than one species of monomer [4b]. Because this is a process-based definition, source-based nomenclature can be easily adapted to the naming of copolymers [10]. The principal problem in this scheme is to define the kind of arrangement in which various types of monomeric units are related to each other. Seven types of separate arrangements have been defined, which are shown in Table i, where A, B and C represent the names of monomers. The monomer names are linked through a connective (infix), such as '-*co*-', to form the name of the copolymer, as in poly(styrene-*co*-acrylonitrile). The order of citation of the monomers is arbitrary.

An equally acceptable alternative scheme, shown in the last column of Table i, utilizes the prefix 'copoly' followed by citation of the names of the monomers used, separated from each other by an oblique stroke. Parentheses are also needed. For example, copoly(styrene/butadiene) denotes an unspecified copolymer of styrene and butadiene. The other connectives of Table i are placed *before* such names, to provide additional structural information, as in *ran*-copoly(ethylene/vinyl acetate), *alt*-copoly(styrene/maleic anhydride). It is not necessary to use parentheses to enclose vinyl acetate, maleic anhydride, methyl acrylate, etc. even though the name of each of

Table i. IUPAC nomenclature of copolymers [10]

Type	Arrangement of monomeric units	Representation	Connective	Example: Name	Example: Alternative name
unspecified	unknown or unspecified	(A-*co*-B)	-*co*-	poly[styrene-*co*-(methyl methacrylate)]	copoly(styrene/methyl methacrylate)
statistical	obeys known statistical laws	(A-*stat*-B-*stat*-C)	-*stat*-	poly(styrene-*stat*-acrylonitrile-*stat*-butadiene)	*stat*-copoly(styrene/acrylonitrile/butadiene)
random	obeys Bernoullian statistics	(A-*ran*-B)	-*ran*-	poly[ethylene-*ran*-(vinyl acetate)]	*ran*-copoly(ethylene/vinyl acetate)
alternating	alternating sequence	$(AB)_n$	-*alt*-	poly[(ethylene glycol)-*alt*-(terephthalic acid)]	*alt*-copoly(ethylene glycol/terephthalic acid)
periodic	periodic with respect to at least three monomeric units	$(ABC)_n$ $(ABB)_n$ $(AABB)_n$ $(ABAC)_n$	-*per*-	poly[formaldehyde-*per*-(ethylene oxide)-*per*-(ethylene oxide)]	*per*-copoly(formaldehyde/ethylene oxide/ethylene oxide)
block	linear arrangement of blocks	—AAAAA—BBBBBB—	-*block*-	polystyrene-*block*-polybutadiene	*block*-copoly(styrene/butadiene)
graft	polymeric side chain different from main chain [a]	—AAAAAAAAAAA— | B B B B B |	-*graft*-	polybutadiene-*graft*-polystyrene	*graft*-copoly(butadiene/styrene)

[a] Main chain (or backbone) is specified first in the name.

these monomers consists of two words; the names of the polymers, as written here, are unambiguous.

The names of copolymers, derived either from the main scheme or the alternative, can be further modified to indicate various structural features. For example, the chemical nature of end groups can be specified as α-X-ω-Y-poly(A-*alt*-B), or as α-butyl-ω-carboxy-*block*-copoly(styrene/butadiene).

Subscripts placed immediately after the formula of the monomeric unit or the block designate the degree of polymerization or repetition. On the other hand, mass and mole fractions and relative molecular masses, which in most cases are average quantities, are expressed by placing corresponding figures after the complete name of the copolymer. The order of citation is as for the monomeric species in the name. Unknown quantities are designated by *a*, *b*, etc. Some examples follow:

1. polybutadiene-*block*-polystyrene (0.75:0.25 *w*)
   block-copoly(butadiene/styrene) (75:25 mass %)

(a block copolymer containing 75 mass % of polybutadiene and 25 mass % of polystyrene);

2. polyisoprene-*graft*-poly(isoprene-*co*-acrylonitrile) (0.85:0.15 *x*)
   graft-copoly[isoprene/co(isoprene/acrylonitrile)] (85:15 mol %)

(a graft copolymer, consisting of a polyisoprene backbone grafted with isoprene and acrylonitrile units in an unspecified arrangement, containing 85 mol % of isoprene units and 15 mol % of acrylonitrile units);

3. polybutadiene-*graft*-polystyrene (75:25 mass %; 90 000:30 000 M_r)

(a graft copolymer consisting of 75 mass % of polybutadiene with a relative molecular mass of 90 000 as the backbone, and 25 mass % of polystyrene in grafted chains with a relative molecular mass of 30 000)

4. *graft*-copoly(butadiene/styrene) (1700:*a* DP)

(a graft copolymer in which the polybutadiene backbone has a degree of polymerization (DP) of 1700 and the polystyrene grafts have an unknown DP).

The complete document [10] should be consulted for the names of more complex copolymers, e.g. those having a multiplicity of grafts or having chains radiating from a central atom.

STRUCTURE-BASED NOMENCLATURE

For organic polymers that are regular (i.e. have only one species of constitutional repeating unit in a single sequential arrangement) and consist only of single strands, the IUPAC has promulgated a structure-based system of naming polymers [6b]. Originally devised by the Polymer Nomenclature Committee of the American Chemical Society [14], the system consists of naming a polymer as 'poly(constitutional repeating unit)', wherein the repeating unit is named as a bivalent organic group according to the usual nomenclature rules for organic chemistry. It is important to note that, in structure-based nomenclature, the name of the constitutional repeating unit has no relationship to the source from which the unit was prepared. The name is simply that of the identifiable constitutional repeating unit in the polymer, and locants for unsaturation, substituents, etc, are dictated by the *structure* of the unit.

The steps involved in naming the constitutional repeating unit are: (1) identification of the unit (taking into account the kinds of atoms in the main chain and the location of substituents); (2) orientation of the unit; and (3) naming of the unit. Examples of names for some common polymers are given in Table ii. Note that, in this system, parentheses are always used to enclose the constitutional repeating unit.

Structure-based nomenclature can be utilized to name polymers of great complexity, provided

Table ii. Examples of structure-based and source-based names for some common polymers [6a]

Structure	Structure-based name	Source-based or trivial name
$\text{-(}CH_2CH_2\text{)}_n\text{-}$	poly(methylene)	polyethylene
$\text{-(}CH(CH_3)CH_2\text{)}_n\text{-}$	poly(propylene)	polypropene
$\text{-(}C(CH_3)_2CH_2\text{)}_n\text{-}$	poly(1,1-dimethylethylene)	polyisobutylene
$\text{-(}C(CH_3){=}CHCH_2CH_2\text{)}_n\text{-}$	poly(1-methyl-1-butenylene)	polyisoprene
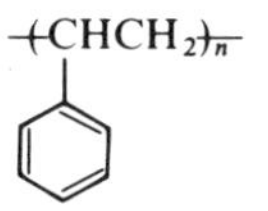	poly(1-phenylethylene)	polystyrene
$\text{-(}CH(Cl)CH_2\text{)}_n\text{-}$	poly(1-chloroethylene)	poly(vinyl chloride)
$\text{-(}CH(CN)CH_2\text{)}_n\text{-}$	poly(1-cyanoethylene)	polyacrylonitrile
$\text{-(}CH(OCOCH_3)CH_2\text{)}_n\text{-}$	poly(1-acetoxyethylene)	poly(vinyl acetate)
$\text{-(}CF_2CH_2\text{)}_n\text{-}$	poly(1,1-difluoroethylene)	poly(vinylidene fluoride)
$\text{-(}CF_2CF_2\text{)}_n\text{-}$	poly(difluoromethylene)	polytetrafluoroethylene
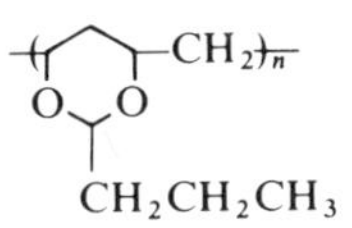	poly[(2-propyl-1,3-dioxane-4,6-diyl)methylene]	poly(vinyl butyral)
$\text{-(}C(CH_3)(COOCH_3)CH_2\text{)}_n\text{-}$	poly[1-(methoxycarbonyl)-1-methylethylene]	poly(methyl methacrylate)
$\text{-(}OCH_2CH_2\text{)}_n\text{-}$	poly(oxyethylene)	poly(ethylene oxide)
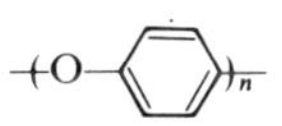	poly(oxy-1,4-phenylene)	poly(phenylene oxide)
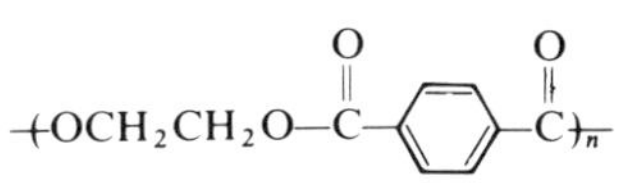	poly(oxyethyleneoxyterephthaloyl)	poly(ethylene terephthalate)
$\text{-(}NH{-}C({=}O)(CH_2)_4C({=}O){-}NH(CH_2)_6\text{)}_n\text{-}$	poly(iminoadipoyliminohexamethylene)	poly(hexamethylenediamine-*co*-adipic acid) *or* poly(hexamethylene adipamide)
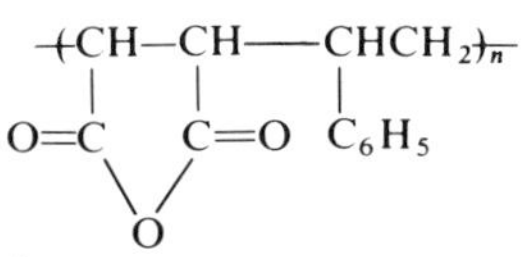	poly[(2,5-dioxotetrahydrofuran-3,4-diyl)-(1-phenylethylene)]	poly(maleic anhydride-*co*-styrene)

only that they be regular and single-stranded. Among these are polymers the constitutional repeating units of which consist, themselves, of a series of smaller subunits, polymers with hetero atoms or heterocyclic ring systems in the main chain, and polymers with substituents on acyclic or cyclic subunits of constitutional repeating units. It should be noted that structure-based nomenclature is also applicable to copolymers having a regular structure, regardless of the starting materials used, e.g. poly(oxyethyleneoxyterephthaloyl) for poly(ethylene terephthalate) [10].

In principle, it should be possible to extend the existing structure-based nomenclature beyond regular, single-strand polymers to reacted polymers, cross-linked polymers, ladder polymers, spiro polymers, and other more complicated systems. Work along these lines is currently under way.

Structure-based nomenclature has gained wide acceptance in the scientific literature, e.g. in *Chemical Abstracts*, because it overcomes many of the deficiencies of source-based nomenclature.

Inorganic and coordination polymers

The nomenclature of regular single-strand inorganic and coordination polymers is governed by the same fundamental principles as that for single-strand organic polymers [8b]. The name of such a polymer is that of the smallest constitutional repeating unit prefixed by the terms 'poly', '*catena*' (for linear chains) or other structural indicator, and designations for end groups. The structural units are named by the nomenclature rules for inorganic and coordination chemistry. Some examples are:

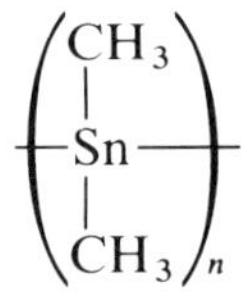

catena-poly[dimethyltin]

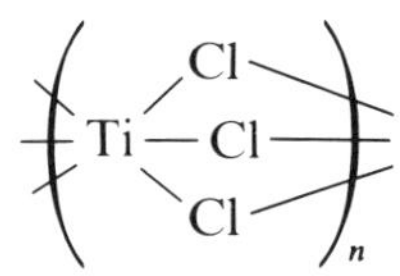

catena-poly[titanium-tri-μ-chloro]

catena-poly[nitrogen-μ-thio]

catena-poly[silver-μ-(cyano-*N*:*C*)]

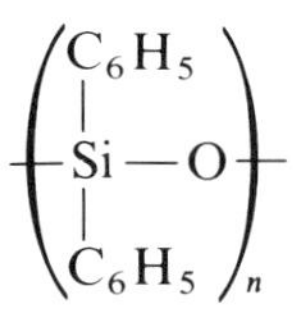

catena-poly[(diphenylsilicon)-μ-oxo]

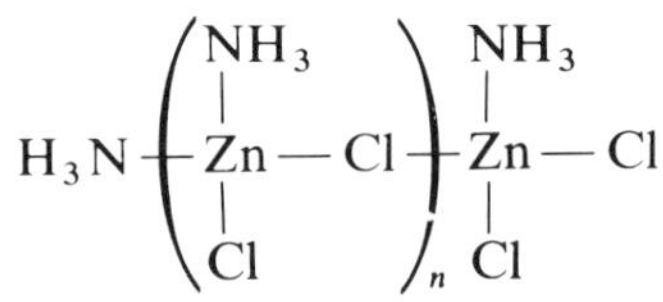

α-ammine-ω-(amminedichlorozinc)-*catena*-poly[(amminechlorozinc)-μ-chloro]

Stereochemical definitions and notations [7b]

Structure-based nomenclature of regular polymers can denote stereochemical features if one utilizes as the repeating unit the configurational base unit, i.e. a constitutional repeating unit having one or more sites of defined stereoisomerism in the main chain of a polymer molecule [4b]. Structure-based names are then derived in the usual fashion. The principal problem is to define the various stereochemical features that are possible in a polymer.

The pioneering work of Natta and co-workers introduced the concept of tacticity, i.e. the orderliness of the succession of configurational repeating units in the main chain of a polymer.

For example, in poly(propylene), possible steric arrangements are (shown in Fischer projections displayed horizontally):

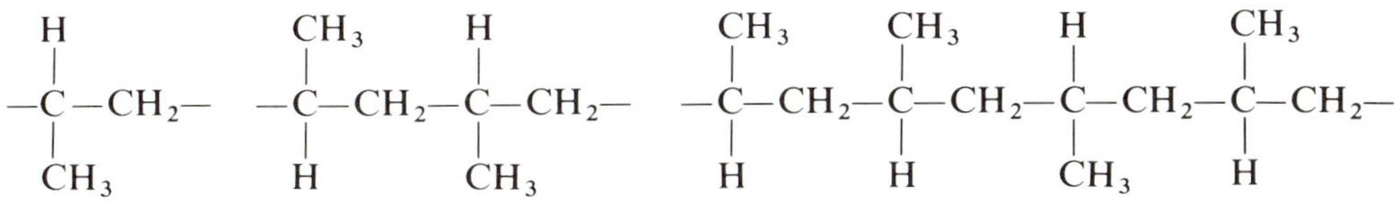

and the corresponding polymers have the following structures:

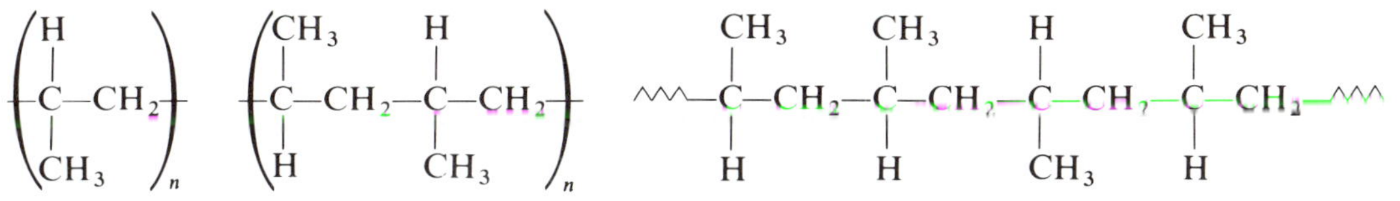

isotactic syndiotactic atactic

The isotactic polymer has only one species of configurational base unit in a single sequential arrangement, the syndiotactic polymer shows an alternation of configurational base units that are enantiomeric, whereas in the atactic polymer the molecules have equal numbers of the possible configurational base units in a random sequence distribution. This can be generalized as follows:

isotactic

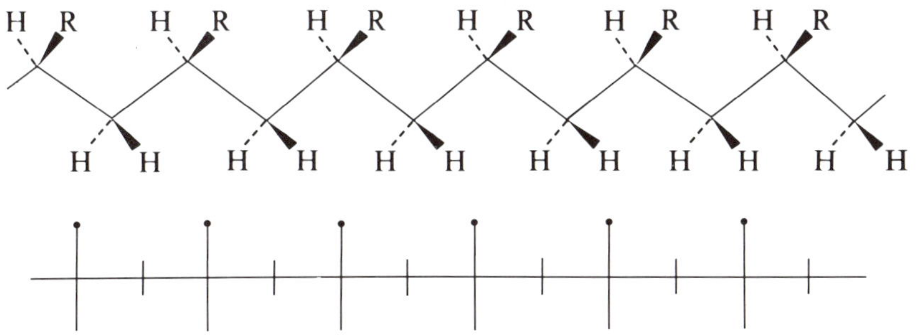

syndiotactic

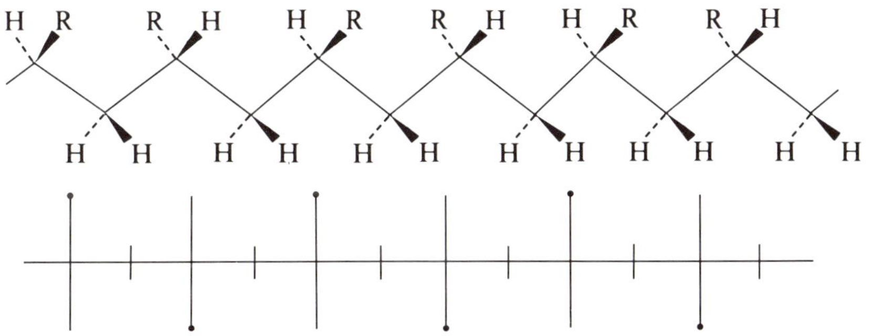

atactic

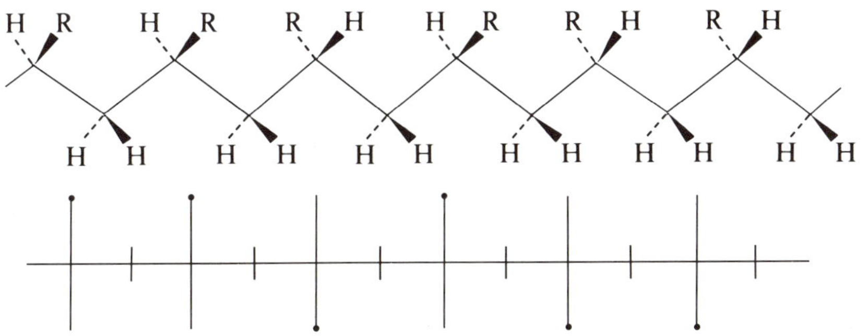

Further examples of tactic polymers are:

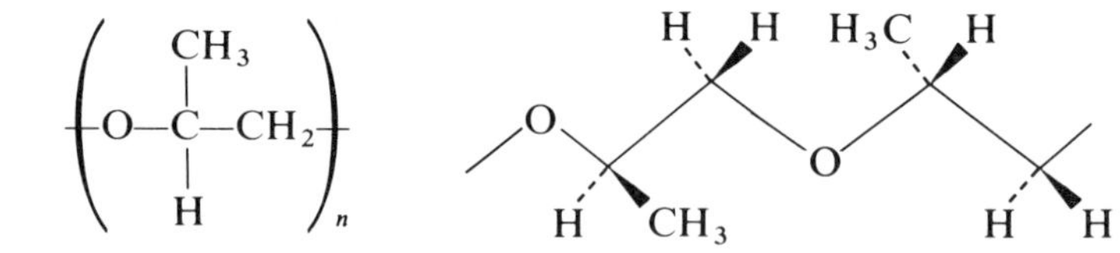

isotactic poly(oxypropylene)

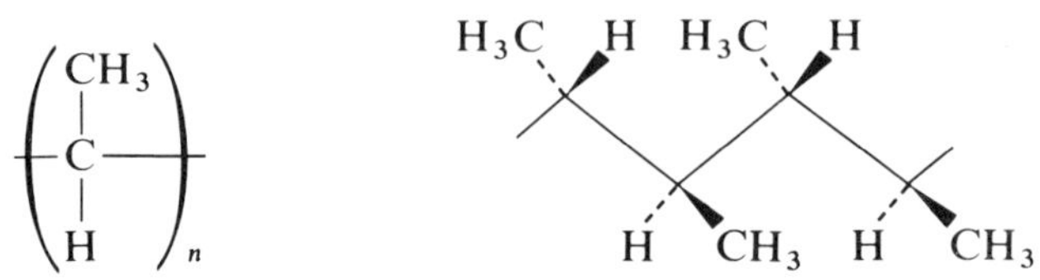

isotactic poly(ethylidene)

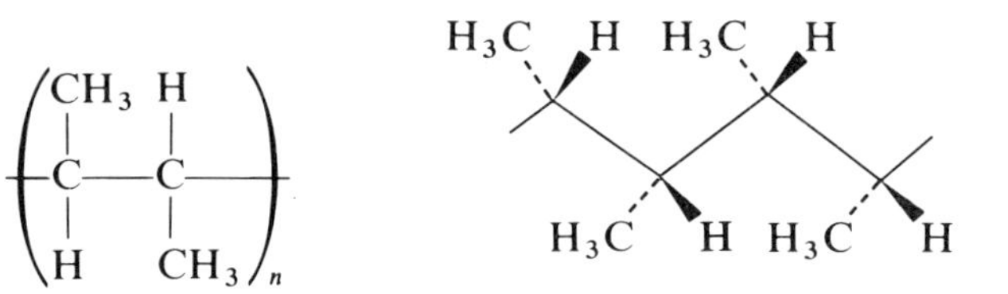

syndiotactic poly(ethylidene)

The concept of a stereoblock is illustrated in the following example of a regular poly(propylene) chain, in which the stereoblocks are denoted by

The sequence of identical relative configurations of adjacent units that characterizes a stereoblock is terminated at each end of the block. The dashed lines

enclose two of many possible configurational sequences, which may or may not be identical with stereoblocks.

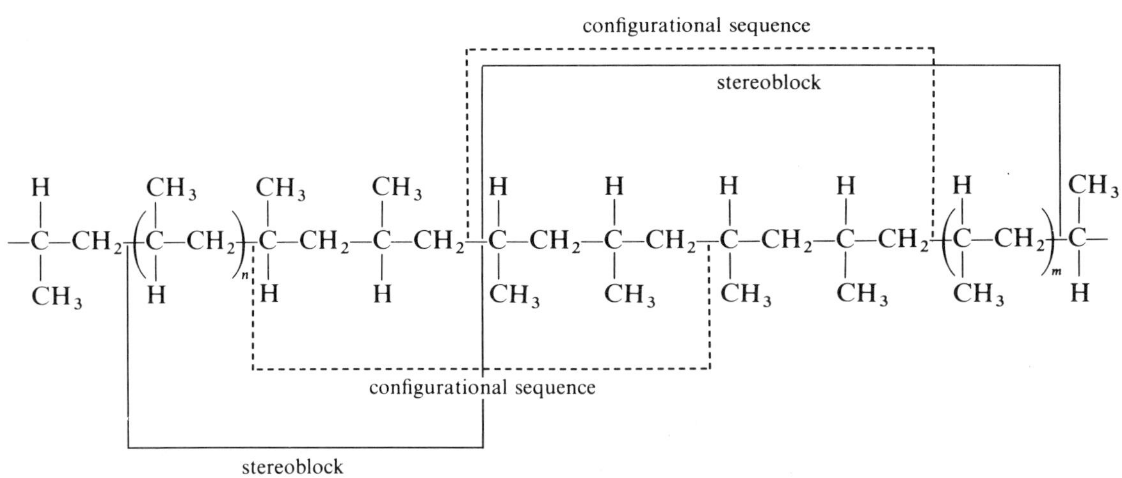

The complete IUPAC document [7b] should always be consulted for complex cases and for the notations used to designate conformations of polymer molecules (bond lengths, bond angles, torsion angles, helix sense, isomorphous and enantiomorphous structures, line repetition groups and symmetry elements, etc) as well as for the various stereochemical definitions.

TRADE NAMES AND ABBREVIATIONS

Because the systematic names of polymers can be cumbersome, trade names and abbreviations are frequently used as a shortcut in industrial literature and oral communication. For example, the simpler generic name nylon-6,6 (where the first number refers to the number of carbon atoms of the diamine and the second number to that of the diacid fragment of the polyamide) appears often in the literature in lieu of the systematic name, poly(iminoadipoyliminohexamethylene). Useful compilations of trade names for polymers can be found in References [15] and [16].

Perhaps the most widely used shortcut is the use of abbreviations for common industrial polymeric materials. The IUPAC recognizes that there may be advantages in some cases to use abbreviations, but urges that each abbreviation be fully defined the first time it appears in the text and that no abbreviation be used in titles of publications. Because there are inherent difficulties in assigning systematic and unique abbreviations to polymeric structures, only a short list has the IUPAC's official sanction [5d]. ISO has published a more extensive list [17], and the American Chemical Society has compiled a master list of all known abbreviations in the polymer field [18].

CONCLUSION

As this chapter has shown, polymer nomenclature has evolved and a substantial body of systematic rules and definitions already exists. The reader is strongly encouraged to consult the full documents for detailed guidance. Recent issues of *Chemical Abstracts* are additional authoritative sources of polymer nomenclature.

NORBERT M. BIKALES

REFERENCES

IUPAC publications

1 Report on nomenclature in the field of macromolecules. *J. Polym. Sci.* **8**, 257–277 (1952). Obsolete.

2a Report on nomenclature dealing with steric regularity in high polymers. *J. Polym. Sci.* **56**, 153–161 (1962). Superseded by Ref. [2b].

2b Report on nomenclature dealing with steric regularity in high polymers. *Pure Appl. Chem.* **12**, 643–656 (1966); *Makromol. Chem.* **82**, 1–15 (1965). Superseded by Ref. [7b].

3 Report of the Committee on Nomenclature of the International Commission on Macromolecules. *J. Polym. Sci., Part B* **6**, 257–260 (1968). Obsolete.

4a Basic definitions of terms relating to polymers. *IUPAC Inf. Bull. Append.* No. **13** (1971). Superseded by Ref. [4b].

4b Basic definitions of terms relating to polymers 1974. *Pure Appl. Chem.* **40**, 477–491 (1974). Reprinted as Chapter 1, this volume.

5a Recommendations for abbreviations of terms relating to plastics and elastomers. *Pure Appl. Chem.* **18**, 581–589 (1969). Obsolete.

5b List of abbreviations for synthetic polymers and polymer materials. *IUPAC Inf. Bull. Append.* No. **12** (1971). Superseded by Ref. [5d].

5c List of standard abbreviations (symbols) for synthetic polymers and polymer materials 1974. *Pure Appl. Chem.* **40**, 473–476 (1974). Superseded by Ref. [5d].

5d Use of abbreviations for names of polymeric substances (recommendations 1986). *Pure Appl. Chem.* **59**, 691–693 (1987). Reprinted as Chapter 9, this volume.

6a Nomenclature of regular single-strand organic polymers. *IUPAC Inf. Bull. Append.* No. **29** (1972);

Macromolecules **6**, 149–158 (1973); *J. Polym. Sci., Polym. Lett. Ed.* **11**, 389–414 (1973). Superseded by Ref. [6b].

6b Nomenclature of regular single-strand organic polymers (rules approved 1975). *Pure Appl. Chem.* **48**, 373–385 (1976). Reprinted as Chapter 5, this volume.

7a Stereochemical definitions and notations relating to polymers (provisional). *Pure Appl. Chem.* **51**, 1101–1121 (1979). Superseded by Ref. [7b].

7b Stereochemical definitions and notations relating to polymers (recommendations 1980). *Pure Appl. Chem.* **53**, 733–752 (1981). Reprinted as Chapter 2, this volume.

8a Nomenclature for regular single-strand and quasi-single-strand inorganic and coordination polymers (provisional). *Pure Appl. Chem.* **53**, 2283–2302 (1981). Superseded by Ref. [8b].

8b Nomenclature for regular single-strand and quasi-single-strand inorganic and coordination polymers (recommendations 1984). *Pure Appl. Chem.* **57**, 149–168 (1985). Reprinted as Chapter 6, this volume.

9 Note on the terminology for molar masses in polymer science. *Makromol. Chem.* **185**, Appendix to No. 1 (1984); *J. Polym. Sci., Polym. Lett. Ed.* **22**, 57 (1984); *J. Colloid Interface Sci.* **101**, 277 (1984); *J. Macromol. Sci., Chem.* **A21**, 903–904 (1984); *Br. Polym. J.* **17**, 92 (1985).

10 Source-based nomenclature for copolymers (recommendations 1985). *Pure Appl. Chem.* **57**, 1427–1440 (1985). Reprinted as Chapter 7, this volume.

11 Definitions of terms relating to individual macromolecules, their assemblies, and dilute polymer solutions (recommendations 1988). *Pure Appl. Chem.* **61**, 211–241 (1989). Reprinted as Chapter 3, this volume.

12 Definitions of terms relating to crystalline polymers (recommendations 1988). *Pure Appl. Chem.* **61**, 769–785 (1989). Reprinted as Chapter 4, this volume.

13 A classification of linear single-strand polymers (recommendations 1988). *Pure Appl. Chem.* **61**, 243–254 (1989). Reprinted as Chapter 8, this volume.

Other publications

14 Nomenclature Committee, Division of Polymer Chemistry, American Chemical Society. A structure-based nomenclature for linear polymers. *Macromolecules* **1**, 193–198 (1968).

15 M. Ash and I. Ash. *Encyclopedia of Plastics, Polymers, and Resins.* Chemical Publishing, New York (1982).

16 H.-G. Elias and R.A. Pethrick (Eds). *Polymer Yearbook.* Harwood Academic Publishers, New York (1984), p. 113.

17 International Organization for Standardization, International Standard ISO 1043–1984. *Plastics—Symbols and Codes.* Part 1: Symbols for basic polymers and their modifications, and for plasticizers.

18 Nomenclature Committee, Division of Polymer Chemistry, American Chemical Society. Abbreviations for thermoplastics, thermosets, fibers, elastomers, and additives. *Polym. News* **9**, 101–110 (1983); **10**, 169–172 (1985).

Definitions

1: Basic Definitions of Terms Relating to Polymers (1974)

PREAMBLE

In the growing field of polymer science, structure-based concepts have received increasing emphasis in the past decades. As a consequence, many of the basic definitions have required refinement.

This report updates and supersedes portions of the 1952 and later reports of the Subcommission on Nomenclature of the IUPAC Commission on Macromolecules [1a, 1b, 1c, 1d] as well as that of the present Commission [2].

The present report is intended to serve the needs of those working in the fundamental areas of polymer science, although the existence of a considerable overlap between the areas of fundamental and applied polymer science is recognized. For definitions in the field of plastics, readers are referred to the ISO Recommendations [3].

For the presentation of clear concepts, idealized definitions are required. At the same time, the realities of polymer science must be faced. Real polymers deviate more or less from ideality on the molecular as well as on the bulk levels. In these definitions, such deviations are neglected. Nevertheless, the definitions given here can be applied as well to the predominating structural features of real polymer molecules.

Two broad sets of definitions are presented. One of these is based on the structure of polymer molecules and the other on the processes by which polymeric substances come into being (Fig. 1.1). The first type of definition is termed 'structure-based' and the second 'process-based'. The primary definition of *polymer* is structure-based. The process-based set of definitions is linked to the primary definition of *polymer* through the definitions of the term *polymerization* and *monomer*. All other definitions are derived from these terms.

DEFINITIONS

Term	*Definition*
1 Primary definitions	
1.1 polymer	A substance composed of molecules characterized by the multiple repetition of one or more species of atoms or groups of atoms (constitutional units, see Definition 1.3) linked to each other in amounts sufficient to provide a set of properties that do not vary

Those served on the Commission during the preparation of this document were K.L. Loening (*Chairman*), P. Corradini, L.C. Cross, R.B. Fox, V.V. Korshak, N.A. Platé, W. Ring, G.J. Smets, C. Suhr and T. Tsuruta. Serving as observers were N.M. Bikales, W.E. Cohn, G. Kline, C. Liébecq and O. Wichterle.
Reprinted from *Pure Appl. Chem.* **40**, 477–491 (1974).

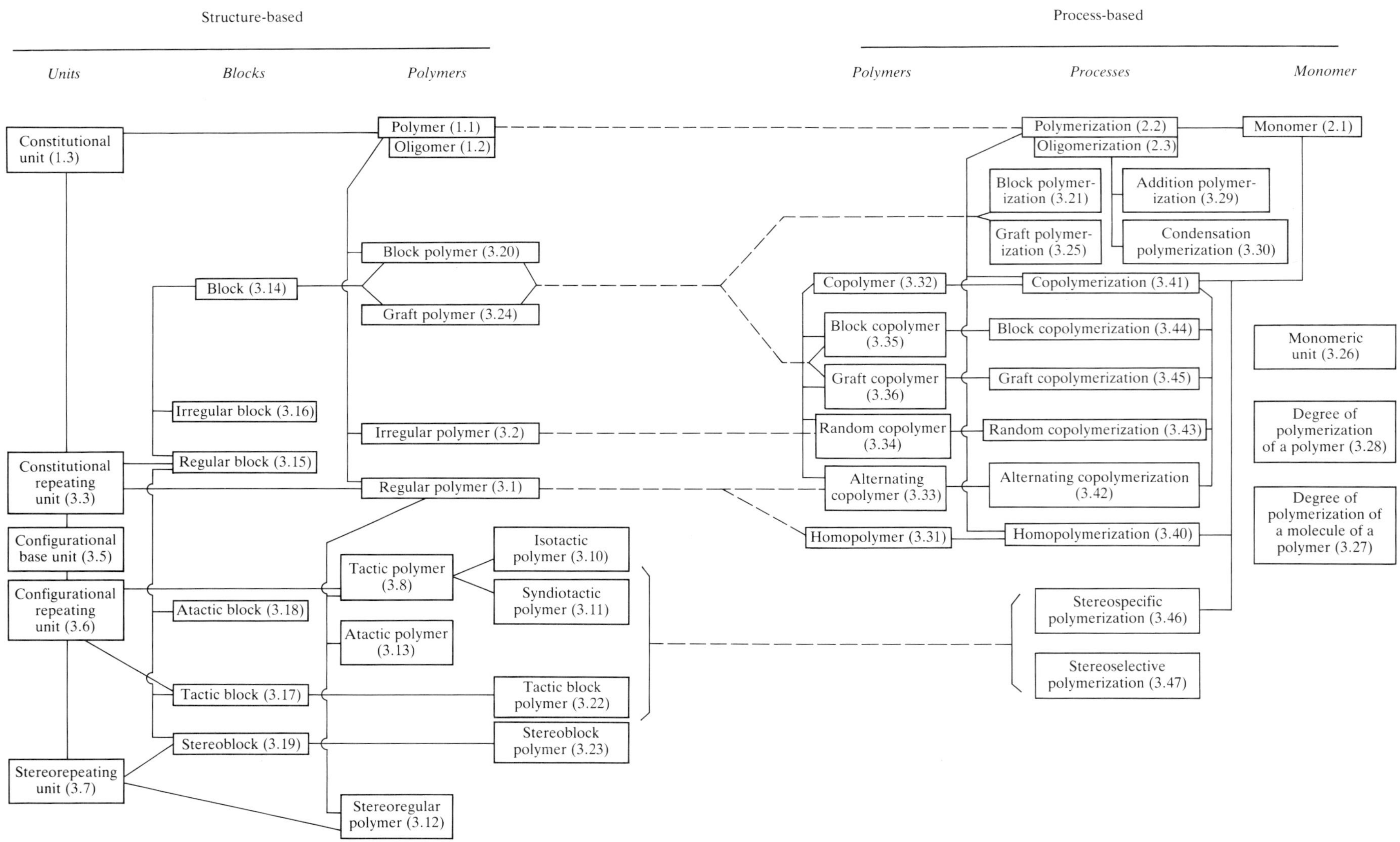

Fig. 1.1 Interrelationships among basic definitions of terms relating to polymers.

Term	*Definition*
	markedly with the addition or removal of one or a few of the constitutional units. [structure-based]
1.2 oligomer	A substance composed of molecules containing a few of one or more species of atoms or groups of atoms (constitutional units) repetitively linked to each other. The physical properties of an oligomer vary with the addition or removal of one or a few of the constitutional units from its molecules. [structure-based]
1.3 constitutional unit	A species of atom or group of atoms present in a chain of a polymer or oligomer molecule. [structure-based]

2 Secondary definitions

2.1 monomer	A compound consisting of molecules each of which can provide one or more constitutional units. [process-based]
2.2 polymerization	The process of converting a monomer or a mixture of monomers into a polymer. [process-based]
2.3 oligomerization	The process of converting a monomer or a mixture of monomers into an oligomer. [process-based]

3 Derived definitions

3.1 regular polymer	A polymer whose molecules can be described by only one species of constitutional unit in a single sequential arrangement. [structure-based]
3.2 irregular polymer	A polymer whose molecules cannot be described by only one species of constitutional unit in a single sequential arrangement. [structure-based]
3.3 constitutional repeating unit	The smallest constitutional unit whose repetition describes a regular polymer. [structure-based]
3.4 configurational unit	A constitutional unit having one or more sites of defined stereoisomerism. [structure-based]

Term	*Definition*
3.5 configurational base unit	A constitutional repeating unit whose configuration is defined at least at one site of stereoisomerism in the main chain of a polymer molecule. *Note* 1: In a regular polymer, a configurational base unit corresponds to the constitutional repeating unit. *Note* 2: Two configurational base units are called enantiomeric when they are mirror images at the plane containing the main-chain bonds. [structure-based]
3.6 configurational repeating unit	The smallest set of one, two or more successive configurational base units that prescribes configurational repetition at one or more sites of stereoisomerism in the main chain of a polymer molecule. [structure-based]
3.7 stereorepeating unit	A configurational repeating unit having defined configuration at all sites of stereoisomerism in the main chain of a polymer molecule. [structure-based]
3.8 tactic polymer	A regular polymer whose molecules can be described by only one species of configurational repeating unit in a single sequential arrangement. [structure-based]
3.9 tacticity	The orderliness of the succession of configurational repeating units in the main chain of a polymer molecule. [structure-based]
3.10 isotactic polymer	A regular polymer whose molecules can be described by only one species of configurational base unit (having chiral or prochiral atoms in the main chain) in a single sequential arrangement. *Note:* In an isotactic polymer molecule the configurational repeating unit is identical with the configurational base unit. [structure-based]
3.11 syndiotactic polymer	A regular polymer whose molecules can be described by alternation of configurational base units that are enantiomeric. *Note:* In a syndiotactic polymer the configurational repeating unit consists of two configurational base units that are enantiomeric. [structure-based]
3.12 stereoregular polymer	A regular polymer whose molecules can be described by only one species of stereorepeating unit in a single sequential arrangement. [structure-based]
3.13 atactic polymer	A regular polymer whose molecules have a random distribution of equal numbers of the possible configurational base units. [structure-based]

Term		*Definition*
3.14	block	A portion of a polymer molecule comprising many constitutional units, that has at least one constitutional or configurational feature not present in the adjacent portions. *Note:* The definitions that relate to polymer can also be applied to block. [structure-based]
3.15	regular block	A block that can be described by only one species of constitutional repeating unit in a single sequential arrangement. [structure-based]
3.16	irregular block	A block that cannot be described by only one species of constitutional repeating unit in a single sequential arrangement. [structure-based]
3.17	tactic block	A regular block that can be described by only one species of configurational repeating unit in a single sequential arrangement. [structure-based]
3.18	atactic block	A regular block that has a random distribution of equal numbers of the possible configurational base units. [structure-based]
3.19	stereoblock	A regular block that can be described by one species of stereorepeating unit in a single sequential arrangement. [structure-based]
3.20	block polymer	A polymer whose molecules consist of blocks connected linearly. The blocks are connected directly or through a constitutional unit that is not part of the blocks. [structure-based]
3.21	block polymerization	Polymerization in which a block polymer is formed. [process-based]
3.22	tactic block polymer	A polymer whose molecules consist of tactic blocks connected linearly. [structure-based]
3.23	stereoblock polymer	A polymer whose molecules consist of stereoblocks connected linearly. [structure-based]
3.24	graft polymer	A polymer whose molecules have one or more species of block connected to the main chain as side chains, these side chains having constitutional or configurational features different from the constitutional units comprising the main chain, exclusive of junction points. [structure-based]

	Term	*Definition*
3.25	graft polymerization	Polymerization in which a graft polymer is formed. [process-based]
3.26	monomeric unit (mer)	The largest constitutional unit contributed by a single monomer molecule in a polymerization process. [process-based]
3.27	degree of polymerization of a molecule of a polymer	The number of monomeric units in a molecule of a polymer. [process-based]
3.28	degree of polymerization of a polymer	The average value of the degree of polymerization of the molecules of a polymer. *Note:* The method of averaging must be stated: for example, number-average degree of polymerization. [process-based]
3.29	addition polymerization	Polymerization by a repeated addition process. [process-based]
3.30	condensation polymerization, polycondensation	Polymerization by a repeated condensation process (i.e. with elimination of simple molecules). [process-based]
3.31	homopolymer	A polymer derived from one species of monomer. [process-based]
3.32	copolymer	A polymer derived from more than one species of monomer. [process-based]
3.33	alternating copolymer	A copolymer in whose molecules two species of monomeric units are distributed in alternating sequence. [process-based]
3.34	random copolymer	A copolymer in whose molecules two or more species of monomeric units are distributed in random sequence. [process-based]
3.35	block copolymer	A block polymer derived from more than one species of monomer. [process-based]
3.36	graft copolymer	A graft polymer derived from more than one species of monomer. [process-based]
3.37	bipolymer	A polymer derived from two species of monomer. [process-based]
3.38	terpolymer	A polymer derived from three species of monomer. [process-based]

Term	*Definition*
3.39 quaterpolymer	A polymer derived from four species of monomer. [process-based]
3.40 homopolymerization	Polymerization in which a homopolymer is formed. [process-based]
3.41 copolymerization	Polymerization in which a copolymer is formed. [process-based]
3.42 alternating copolymerization	Polymerization in which an alternating copolymer is formed. [process-based]
3.43 random copolymerization	Polymerization in which a random copolymer is formed. [process-based]
3.44 block copolymerization	Polymerization in which a block copolymer is formed. [process-based]
3.45 graft copolymerization	Polymerization in which a graft copolymer is formed. [process-based]
3.46 stereospecific polymerization	Polymerization in which a tactic polymer is formed. [process-based]
3.47 stereoselective polymerization	Polymerization in which a polymer molecule is formed from a mixture of stereoisomeric monomer molecules by incorporation of only one stereoisomeric species. [process-based]

EXAMPLES

Constitutional unit

The polymer chain (1)

(1)

can provide the *constitutional units**

*The convention of orienting polymer structures (and the corresponding constitutional and configurational units) from left to right is used in this document. Thus, the bracketted constitutional units in

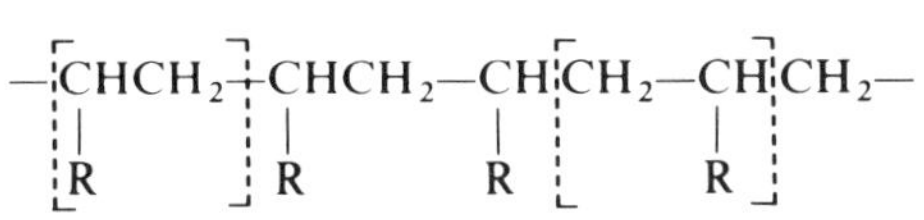

are regarded as different, even though their repetition leads to the same regular polymer.

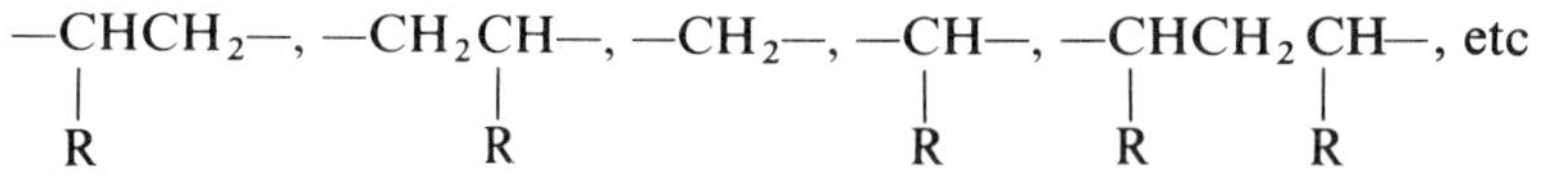

(1a) (1b)

Only the first two constitutional units (1a and 1b) are the smallest ones completely describing this polymer chain.

Constitutional repeating unit

Regular polymer
Either of the constitutional units (1a) or (1b) is a *constitutional repeating unit* and the polymer whose molecules can be described by (1) is a *regular polymer*.

Irregular polymer
The fragment (2)

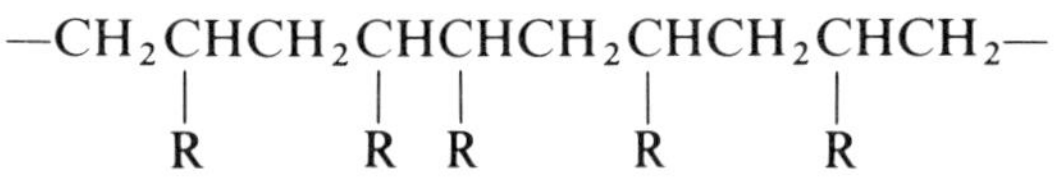

(2)

of a polymer molecule cannot be described by only one constitutional unit, such as (1a) or (1b), in a single sequential arrangement. The polymer whose molecules consist of a random arrangement of (1a) and (1b), as in (2) is an *irregular polymer*.

A polymer whose molecules consist of a random arrangement of

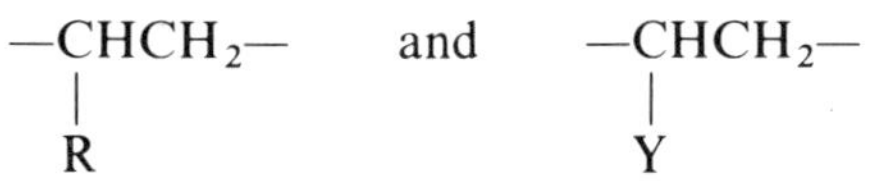

units provides fragments like (3):

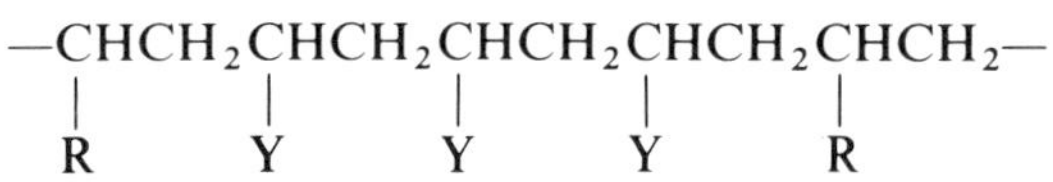

(3)

In this fragment there are the constitutional units

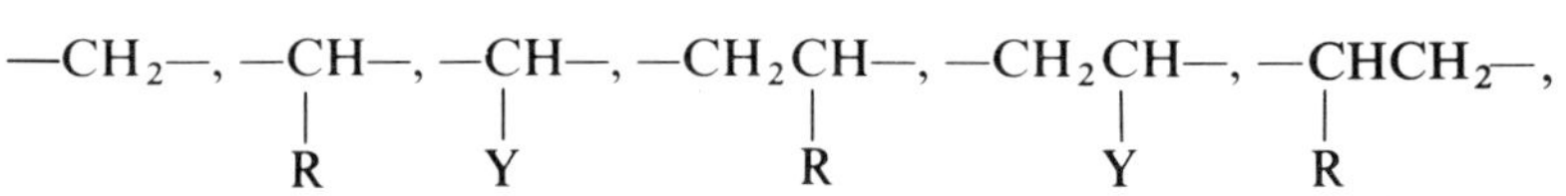

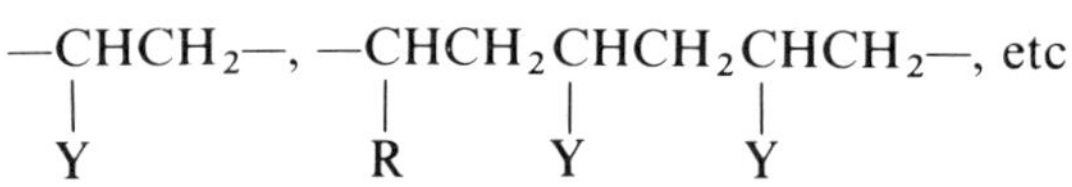

The constitution of (3) cannot be described by any one of these constitutional units alone. Therefore, the polymer molecules corresponding to (3) are those of an irregular polymer.

Monomer
The *monomer* molecule $CH_2{=}CHCH{=}CH_2$ can provide the constitutional units:

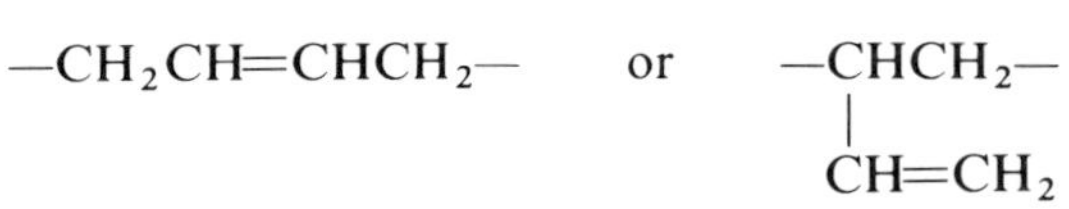

The monomer molecule CH_2N_2 can provide the constitutional unit

$-CH_2-$. The monomer molecule 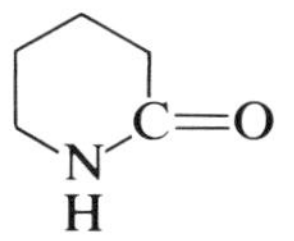 can provide the constitutional unit

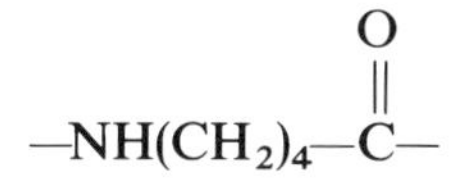

By a condensation process, the monomer molecules $H_2N(CH_2)_6NH_2$ and $ClCO(CH_2)_4COCl$ can lead to the regular polymer molecule

$-[NH(CH_2)_6NHCO(CH_2)_4CO]_n-$

which contains the constitutional units

$-NH(CH_2)_6NH-$ and $-CO(CH_2)_4CO-$

Configurational base unit
In the regular polymer molecule

$-[CH(CH_3)CH_2]_n-$ poly(propylene)

the constitutional repeating unit is $-CH(CH_3)CH_2-$ and the configurational base units are*

These two configurational base units are enantiomeric to each other.

Stereorepeating unit
In a stereoregular poly(propylene), possible *stereorepeating units* are:

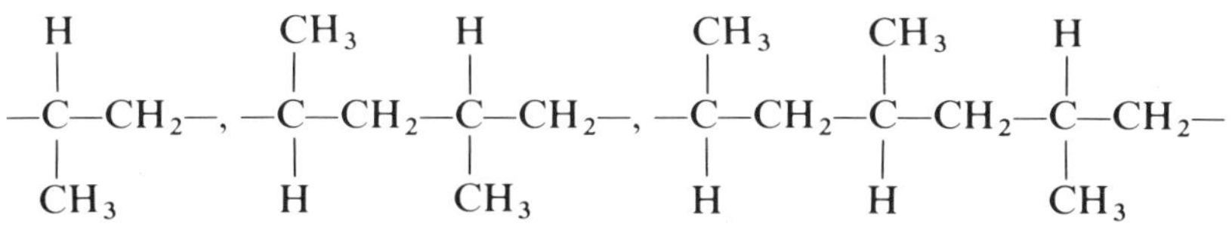

Stereoregular polymer
The corresponding *stereoregular polymers* are:

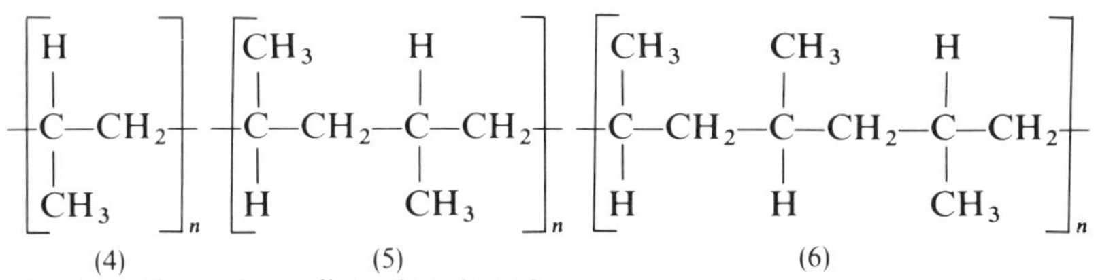

(4) (an isotactic polymer) (5) (a syndiotactic polymer) (6)

*In the drawing of configurations horizontal bonds are below the plane of the paper and vertical bonds above [See IUPAC Rules on Fundamental Stereochemistry, *J. Org. Chem.* **35**, 2849 (1970), Rule E-7.1, note 2]. The main-chain bonds of polymer molecules are drawn as horizontal bonds. Unless otherwise stated the drawings of a configurational base unit, configurational repeating unit, stereorepeating unit, etc provide information concerning *relative* configurations.

The absence in a formula of any one of the horizontal and/or vertical bonds at a chiral or prochiral carbon atom, or of the cis and trans designations, means lack of knowledge about the configuration of the corresponding stereoisomeric centre.

Configurational repeating unit

Tactic polymer, isotactic polymer, syndiotactic polymer

In the polymer $-\!\!\left[CH(COOR)CH(CH_3)\right]_n\!\!-$, if only one main-chain stereoisomeric site of each constitutional repeating unit is defined, as in (7):

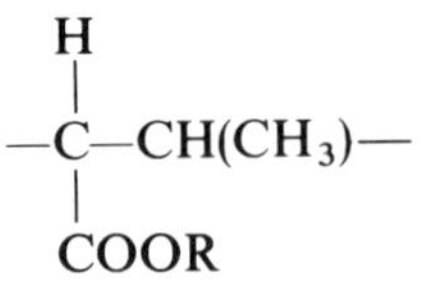

(7)

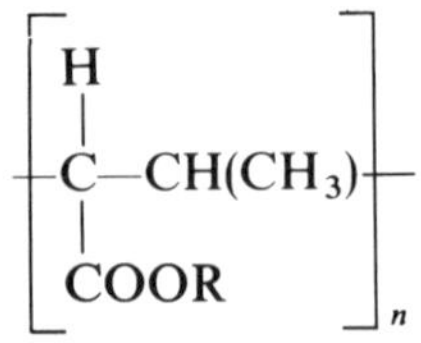

(8)

then (7) is a *configurational repeating unit* and the corresponding polymer (8) is a *tactic* (*isotactic*) *polymer*. Another example of a configurational repeating unit is (9):

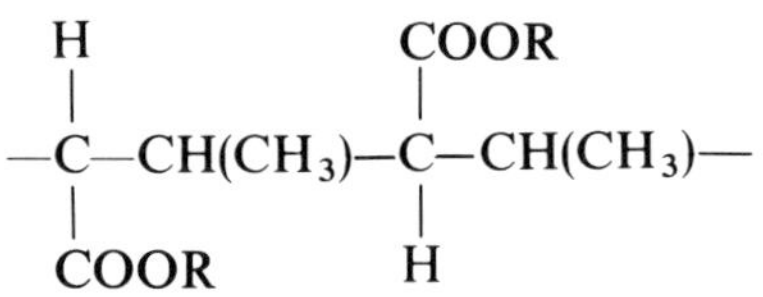

(9)

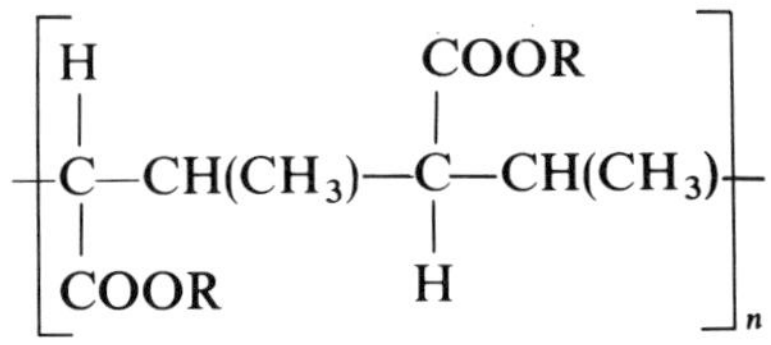

(10)

and the corresponding polymer (10) is a *tactic* (*syndiotactic*) *polymer*. Examples (8) and (10) are not stereoregular because the configuration at the stereoisomeric centre $-CH(CH_3)-$ is not defined. Similarly, (11) and

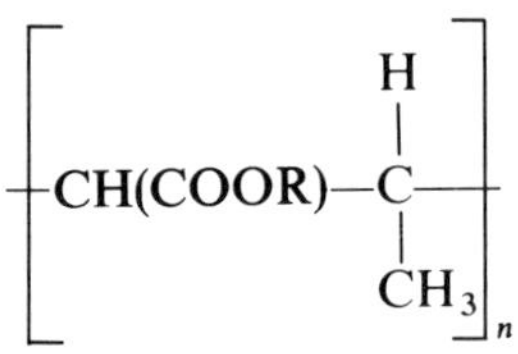

(11)

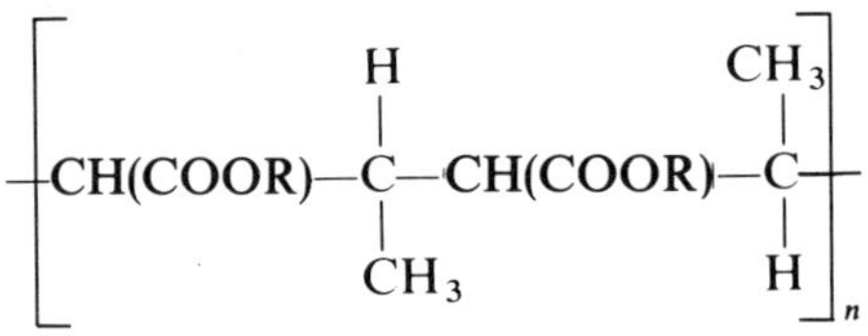

(12)

(12) are not stereoregular because of lack of knowledge about the stereoisomeric centre $-CH(COOR)-$. Examples (4), (5), (6), (8), (10), (11) and (12) are *tactic polymers.* A stereoregular polymer is always a tactic polymer, but a tactic polymer is not always stereoregular, since a tactic polymer need not have every site of stereoisomerism defined.

Atactic polymer

A regular polymer that is *atactic* contains a *random* distribution of equal numbers of the possible configurational base units corresponding to the constitutional repeating unit. Some examples are:

constitutional repeating unit	configurational base units
$-CH(CH_3)CH_2-$	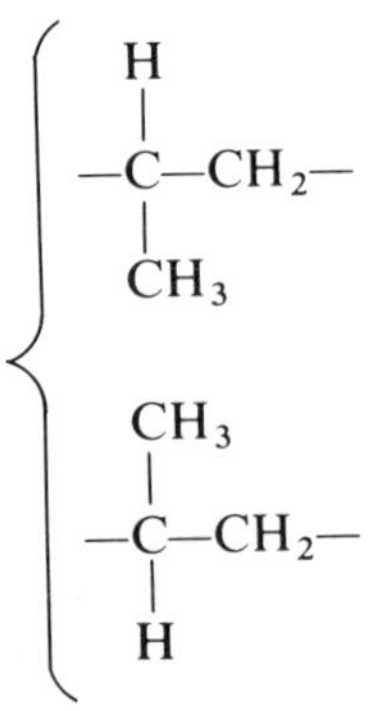

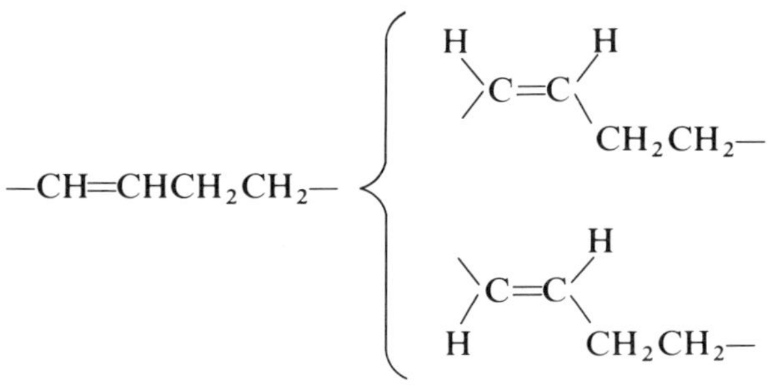

Block

Regular block, block polymer
In the polymer molecule

$$A_k—B_l—A_m—B_n \qquad (13)$$

A_k, B_l, A_m and B_n are *blocks*, and the individual blocks are *regular*. In the *block polymer* molecule (13), A and B may be

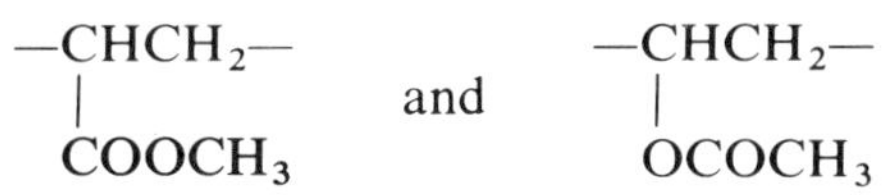

(14) (15)

Block copolymer
The block polymer whose molecules consist of (14) and (15) is a *block copolymer* because (14) and (15) arise from different monomer species.

Tactic block
In the block polymer molecule (13) A and B may be:

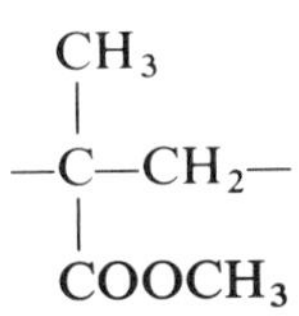
(16)

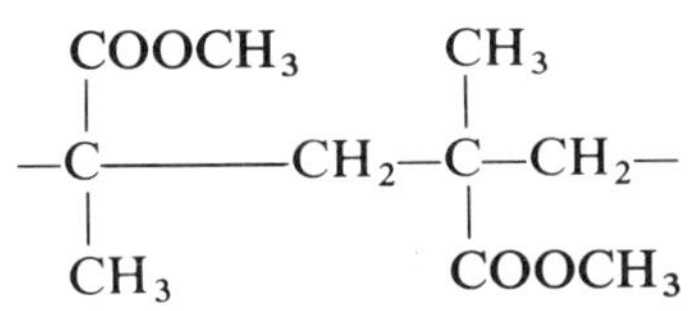
(17)

Stereoblock
The blocks are *stereoblocks*. The block polymer whose molecules consist of (16) and (17) is not a block copolymer because (16) and (17) arise from the same monomer species.

Graft polymer

Graft copolymer
In the *graft polymer* molecule

the A-chain, B_m and B_n are regular blocks, the A-chain is the main chain, and B_m and B_n are the side-chain grafts. The —A— units are junction points and are considered to be part of the main chain. Where A and B are derived from the same monomer, as in

$$-\mathrm{CH{=}CHCH_2CH_2}- \qquad \begin{array}{l} -\mathrm{CH_2CH}- \\ \qquad\ \ | \\ \qquad\ \mathrm{CH{=}CH_2} \end{array}$$

(A) (B)

the polymer is a graft polymer. The graft polymer whose molecules consist of A = (14) and B = (15) is a *graft copolymer.*

REFERENCES

1a IUPAC. Report on nomenclature in the field of macromolecules. *J. Polym. Sci.* **8**, 257–277 (1952).

1b M.L. Huggins, G. Natta, V. Desreux and H. Mark. Report on nomenclature dealing with steric regularity in high polymers. *J. Polym. Sci.* **56**, 153–161 (1962).

1c M.L. Huggins, G. Natta, V. Desreux and H. Mark. Report on nomenclature dealing with steric regularity in high polymers. *Makromol. Chem.* **82**, 1–15 (1965); *Pure Appl. Chem.* **12**, 643–656 (1966).

1d M.L. Huggins, P. Corradini, V. Desreux, O. Kratky and H. Mark. Report of the Committee on Nomenclature of the International Commission on Macromolecules. *J. Polym. Sci., Part B: Polym. Lett.* **6**, 257–260 (1968).

2 Basic definitions of terms relating to polymers. *IUPAC Inf. Bull. Append.* No. **13** (February 1971).

3 ISO Recommendation R 472 (1969).

2: Stereochemical Definitions and Notations Relating to Polymers (1980)

PREAMBLE

A report entitled 'Nomenclature Dealing with Steric Regularity in High Polymers' was issued in 1962 by the Committee on Nomenclature of the Commission on Macromolecules of the IUPAC [1]. Since then, the development of increasingly sophisticated techniques for structure determination has greatly enlarged the field of polymer stereochemistry and this, in turn, has revealed the need for a detailed knowledge of molecular conformations in order to correlate chemical structure with physical properties. The nomenclature relating to the constitution and configuration of macromolecules has been refined, using structure-based concepts, in documents of this Commission [2, 3], while an IUPAC paper [4] on the stereochemistry of organic molecules and an IUPAC-IUB document [5] on abbreviations and symbols to be used for the description of the conformations of polypeptide chains have appeared as definitive publications. The present statement is intended to bring up to date the nomenclature of features corresponding to stereoregularity in polymers; it employs the definitions prescribed in [2] and takes into account all the previously elaborated material cited above; it also introduces new concepts dealing with the microstructure of polymer chains, and it proposes a set of definitions and notations for the description of the conformations of polymer molecules. Consistency with documents [4] and [5] has been maintained as far as is possible.

Throughout this document, stereochemical formulae for polymer chains are shown as Fischer projections rotated through 90°, i.e. displayed horizontally rather than vertically, (at variance with [1]) or as hypothetical extended zigzag chains; the latter occasionally give a clearer indication of the three-dimensional arrangement. It is preferred that the hypothetical extended zigzag chains be consistently drawn with the backbone bond on the extreme left of the formula presented rising from left to right and with the interrupted line, on any given backbone carbon atom, drawn to the left of the full line.

The use of rotated Fischer projections corresponds to the common practice of using horizontal lines to denote polymer backbone bonds, but it is most important to note that this does not give an immediately visual impression of the zigzag chain. In the projections as used in this document, *at each individual backbone carbon atom* the horizontal lines represent bonds directed below the plane of the paper from the carbon atom while the vertical lines project above

Prepared for publication by A.D. Jenkins (UK).
Reprinted from *Pure Appl. Chem.* **53**, 733–752 (1981).

the plane of the paper from the carbon atom. Thus, the rotated Fischer projection

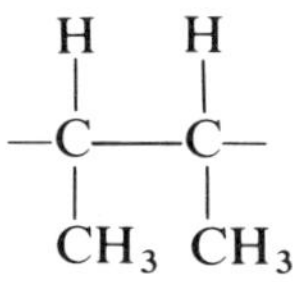 corresponds to 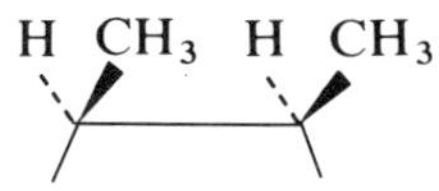 and hence to the zigzag chain: 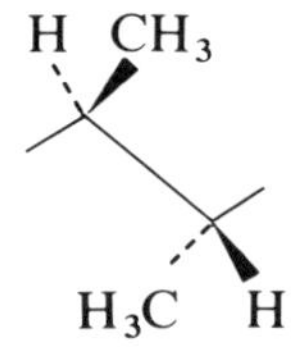

while 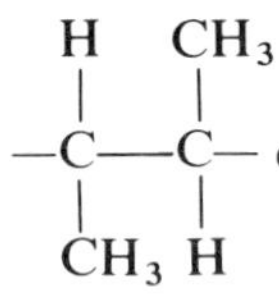 corresponds to 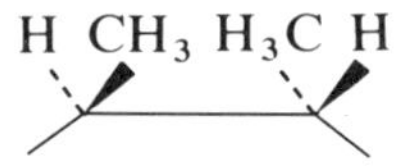 and hence to: 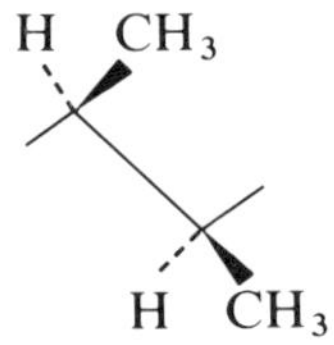

Unless otherwise stated, the drawings of configurational base units, configurational repeating units, stereorepeating units, etc provide information regarding *relative* configurations.

In a polymer molecule, the two portions of the main chain attached to any constitutional unit are, in general, non-identical; consequently, a backbone carbon atom that also bears two different side groups is considered to be a chiral centre.

The absence from a formula of any one of the horizontal or vertical lines at a chiral or prochiral carbon atom (as in examples on pages 30 and 34), or of cis or trans designations at double bonds, indicates that the configuration of that stereoisomeric centre is not known. Also, as in our previous document [2], the convention of orienting polymer structures (and the corresponding constitutional and configurational units) from left to right is used. Thus, the two bracketted constitutional units in

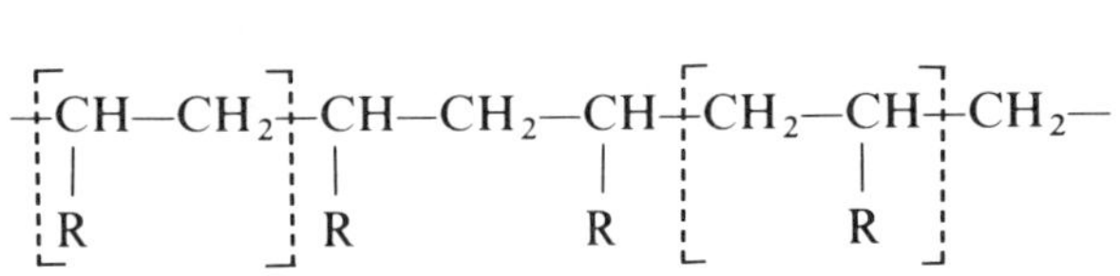

are regarded as different, even though the repetition of either one of them would give the same regular polymer. Some of the definitions presented also appear in a previous paper of this Commission [2], but they are repeated here (with minor grammatical improvement) in order to provide a complete set of stereochemical definitions in a single document.

In order to present clear concepts it is necessary that idealized definitions be adopted but it is recognized that the realities of polymer science must be faced. Deviations from ideality arise with polymers at both molecular and bulk levels in ways that have no parallel with the ordinary small molecules of organic or inorganic chemistry. Although such deviations are not explicitly taken into account in the definitions below, the nomenclature recommended can usefully be applied to the predominant structural features of real polymer molecules, if necessary with self-explanatory, if imprecise, qualifications such as 'almost completely isotactic' or 'highly syndiotactic'. Although such expressions lack the rigour beloved by the purist, every experienced polymer scientist knows that communication in this discipline is impossible without them.

1 BASIC DEFINITIONS

Term	*Definition*
1.1 configurational unit	A constitutional unit having one or more sites of defined stereoisomerism.

Term	*Definition*
1.2 configurational base unit	A constitutional repeating unit, the configuration of which is defined at one or more sites of stereoisomerism in the main chain of a polymer molecule.

In a regular polymer, a configurational base unit corresponds to the constitutional repeating unit.

1.3 configurational repeating unit	The smallest set of one, two or more successive configurational base units that prescribes configurational repetition at one or more sites of stereoisomerism in the main chain of a polymer molecule.
1.4 stereorepeating unit	A configurational repeating unit having defined configuration at all sites of stereoisomerism in the main chain of a polymer molecule.

Note on enantiomeric and diastereoisomeric units
Two configurational units (Definitions 1.1, 1.2, 1.3, 1.4 above) that correspond to the same constitutional unit are considered to be *enantiomeric* if they are non-superposable mirror images. Two non-superposable configurational units that correspond to the same constitutional unit are considered to be *diastereoisomeric* if they are *not* mirror images.

Examples
In the regular polymer molecule $-\!\!\left[CH(CH_3)-CH_2\right]_n-$, poly(propylene), the constitutional repeating unit is $-CH(CH_3)-CH_2-$ and the corresponding configurational base units are

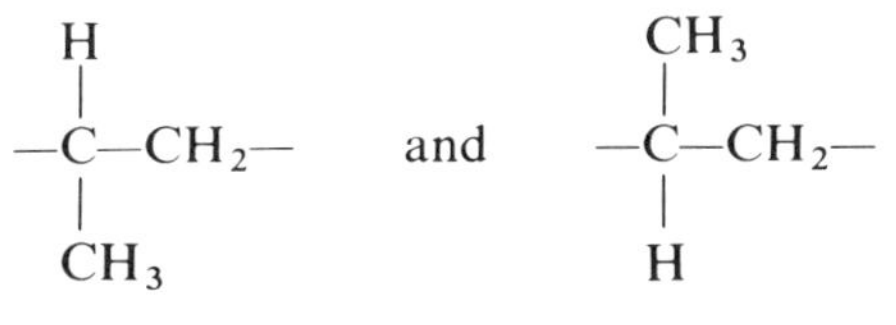

(1) (2)

The configurational base units (1) and (2) are enantiomeric, while the configurational units (1) and (3) cannot be enantiomeric because the constitutional units are different species, according to this nomenclature.

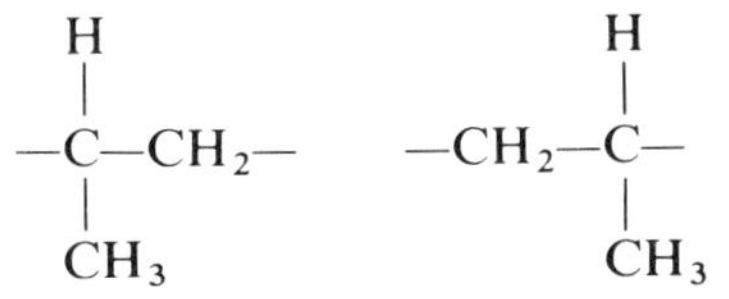

(1) (3)

It is immaterial whether (1) or (2) is taken as the configurational repeating unit and stereorepeating unit of isotactic poly(propylene) (see Definition 1.7); this is so because the two infinite chains, one built up of identical configurational units (1) and the other built up of identical configurational units (2), are *not* enantiomeric and differ only in the chain orientation. Within each pair of units,

such as 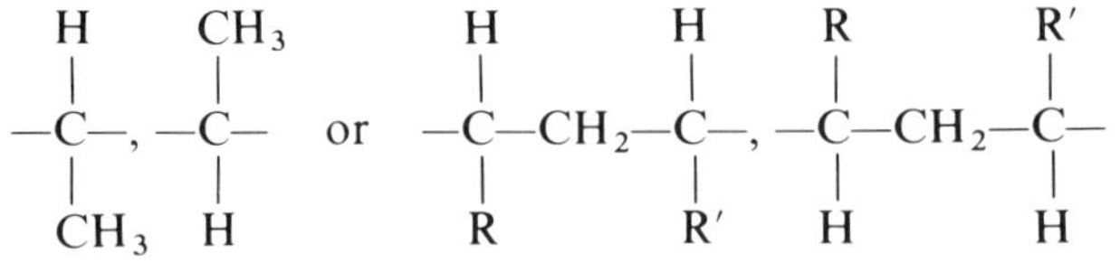

the components are enantiomeric since they are non-superposable mirror images, as defined above. However, with the constitutional unit —CHR—CHR′—CH_2—CH_2—, the two corresponding configurational units

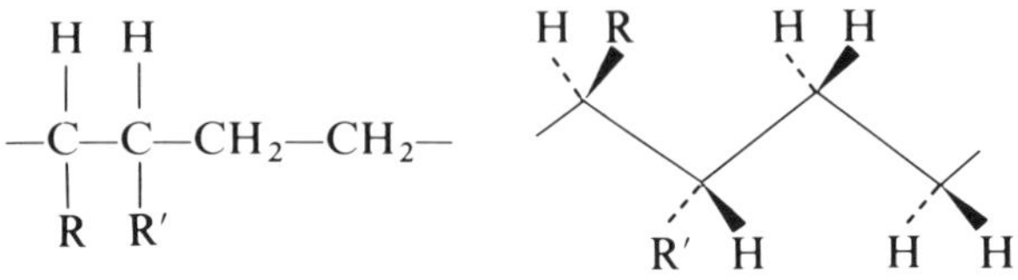

and 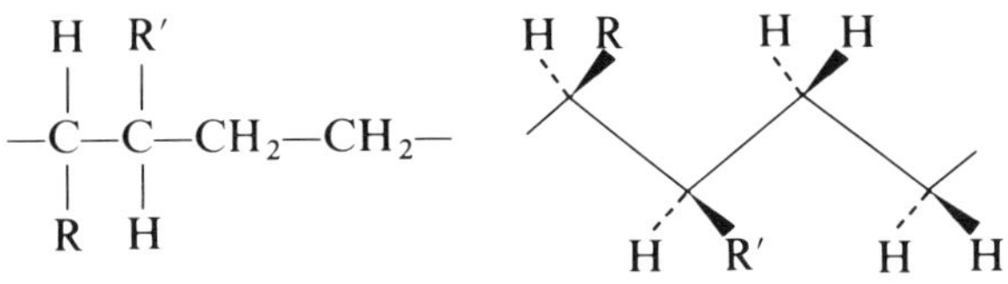

are diastereoisomeric. The units

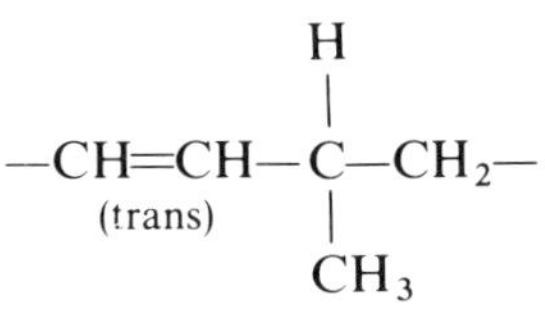 and 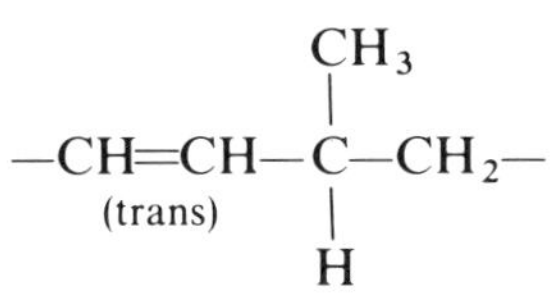

are enantiomeric, while the units

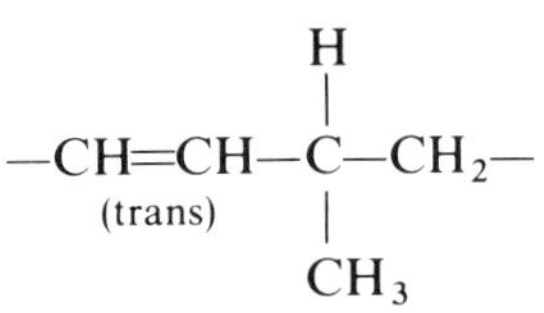 and 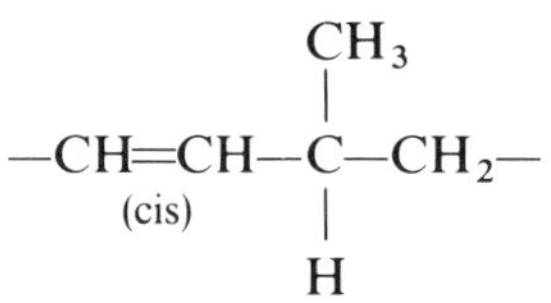

are not enantiomeric, but diastereoisomeric. The simplest possible stereorepeating units in a stereoregular poly(propylene) are

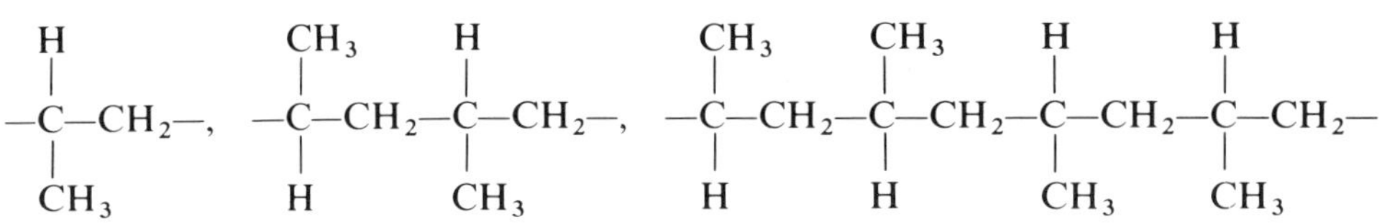

and the corresponding stereoregular polymers are

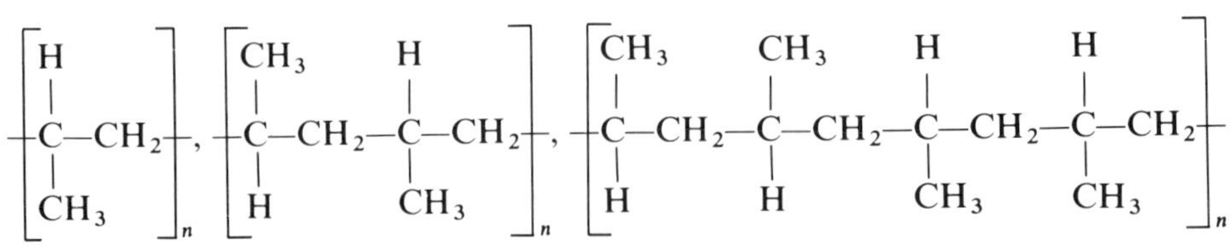

(4) an isotactic polymer, see Definition 1.7

(5) a syndiotactic polymer, see Definition 1.8

(6) a hypothetical heterotactic polymer, see Section 2.2

Term	*Definition*
1.5 tactic polymer	A regular polymer, the molecules of which can be described in terms of only one species of configurational repeating unit in a single sequential arrangement.
1.6 tacticity	The orderliness of the succession of configurational repeating units in the main chain of a polymer molecule. (For the definition of degree of tacticity, see Section 4.)
1.7 isotactic polymer	A regular polymer, the molecules of which can be described in terms of only one species of configurational base unit (having chiral or prochiral atoms in the main chain) in a single sequential arrangement. *Note* In an isotactic polymer, the configurational repeating unit is identical with the configurational base unit.
1.8 syndiotactic polymer	A regular polymer, the molecules of which can be described in terms of alternation of configurational base units that are enantiomeric. *Note* In a syndiotactic polymer, the configurational repeating unit consists of two configurational base units that are enantiomeric.
1.9 stereoregular polymer	A regular polymer, the molecules of which can be described in terms of only one species of stereorepeating unit in a single sequential arrangement.
1.10 atactic polymer	A regular polymer, the molecules of which have equal numbers of the possible configurational base units in a random sequence distribution.

Examples
For the polymer $\left[CH(COOR)CH(CH_3) \right]_n$, if only the ester-bearing main-chain site in each constitutional repeating unit has defined stereochemistry, the configurational repeating unit is (7) and the corresponding isotactic polymer is (8).

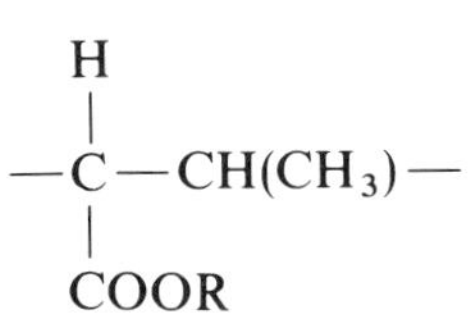

(7)

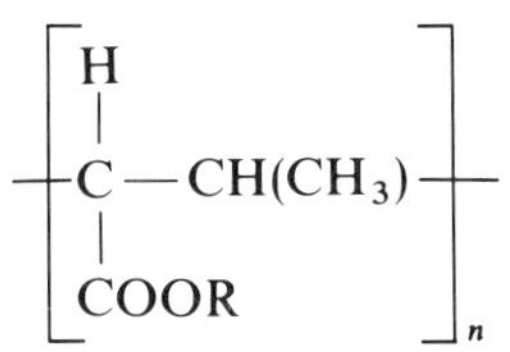

(8)

In the corresponding syndiotactic case, the configurational repeating unit is (9) and the syndiotactic polymer is (10):

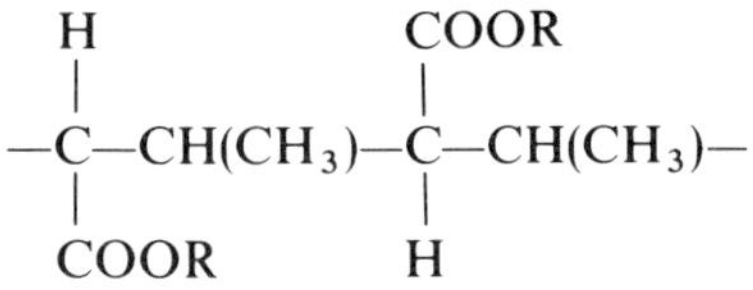

(9)

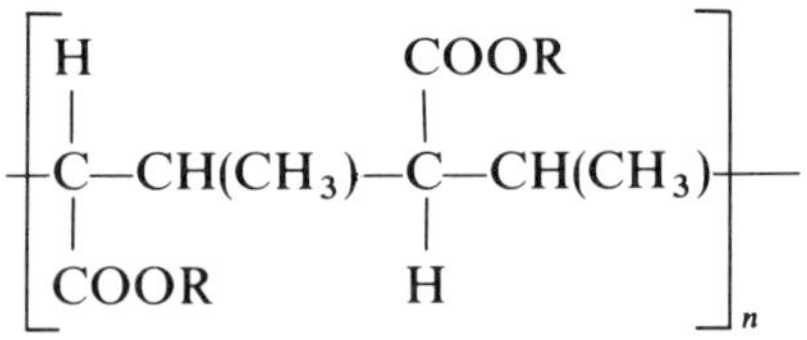

(10)

As the definition of a stereoregular polymer (see Definitions 1.4 and 1.9) requires that the configuration be defined at *all* sites of stereoisomerism, structures (8) and (10) do *not* represent stereoregular polymers. The same is true of (11) and (12), which differ from (8) and (10) in that the sites of specified and unspecified configuration have been interchanged.

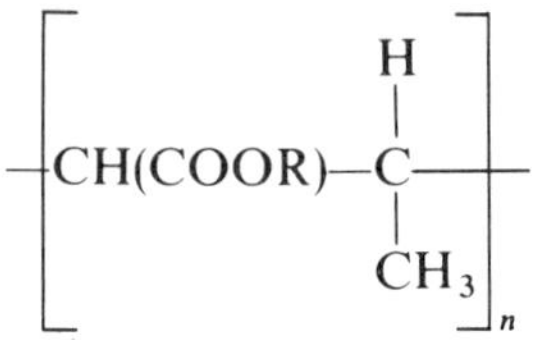

(11)

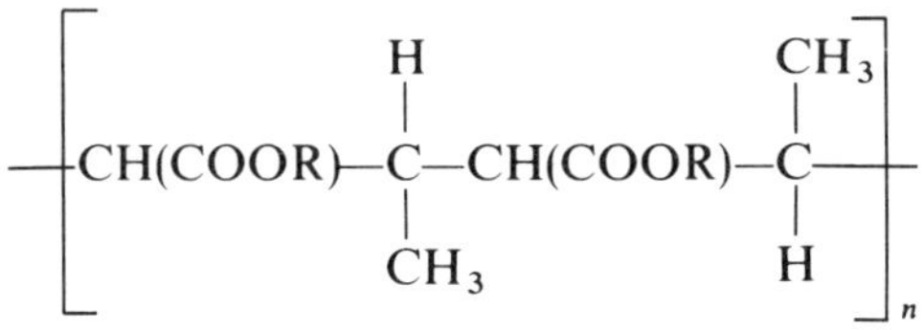

(12)

Examples (4), (5), (6), (8), (10), (11) and (12) are tactic polymers. A stereoregular polymer is always a tactic polymer, but a tactic polymer is not always stereoregular because a tactic polymer need not have *all* sites of stereoisomerism defined.

Further examples of tactic polymers are:

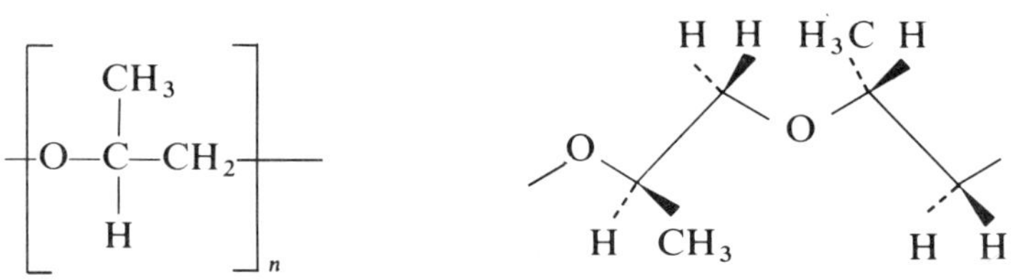

isotactic poly(oxypropylene)

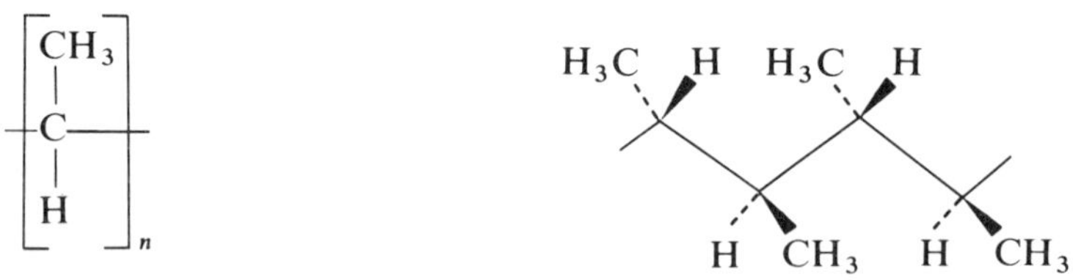

isotactic poly(ethylidene)

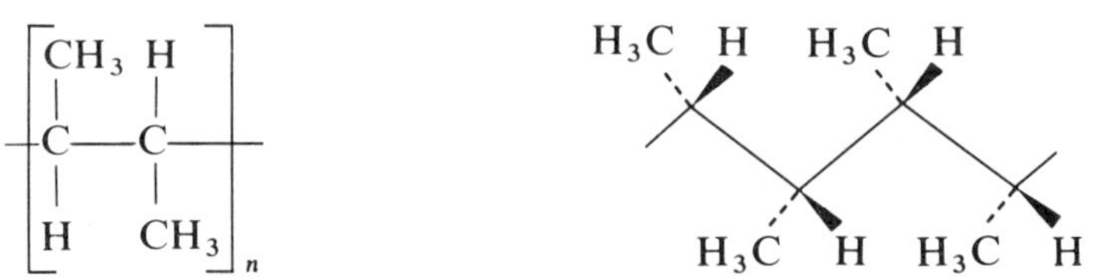

syndiotactic poly(ethylidene)

Note

Structure-based names of tactic polymers are formed before the application of adjectives designating tacticity; thus, 'syndiotactic poly(ethylidene)' is preferred to 'syndiotactic poly(1,2-dimethylethylene)' because a shorter repeating unit is identified, in conformity with the rules in Reference [3].

Note on atactic polymers

As the definition above indicates, a regular polymer, the configurational base units of which contain one site of stereoisomerism only, is atactic if it has equal numbers of the possible types of configurational base units arranged in a random distribution. If the constitutional repeating unit contains more than one site of stereoisomerism, the polymer may be atactic with respect to only one type of site if there are equal numbers of the possible configurations of that site arranged in a random distribution.

Examples

constitutional repeating unit	configurational base units (randomly distributed in an atactic polymer)
$—CH(CH_3)CH_2—$	
$—CH{=}CHCH_2CH_2—$	

A polymer such as $\text{-[}CH{=}CH—CH(CH_3)—CH_2\text{]-}_n$, which has two main-chain sites of stereoisomerism, may be atactic with respect to the double bond only, with respect to the chiral atom only or with respect to both centres of stereoisomerism. If there is a random distribution of equal numbers of units in which the double bond is cis and trans, the polymer is atactic with respect to the double bond, and if there is a random distribution of equal numbers of units containing the chiral atom in the two possible configurations, the polymer is atactic with respect to the chiral atom.* The polymer is completely atactic when it contains, in a random distribution, equal numbers of the four possible configurational base units which have defined stereochemistry at both sites of stereoisomerism.

In addition to isotactic, syndiotactic and atactic polymers (and other well-defined types of tactic polymers), there exists the whole range of possible arrangements between the completely ordered and the completely random distributions of configurational base units, and it is necessary to employ the concept of degree of tacticity (see Section 4) to describe such systems.

*With regard to isomerism about double bonds, it is recommended that the *E* and *Z* designations [4] be used, where appropriate, in describing side-chain configurations and in the names of monomers used in source-based polymer nomenclature. In structure-based polymer names and in descriptions of configuration about double bonds in polymer main chains, the use of cis and trans is preferred.

Term		*Definition*
1.11	stereospecific polymerization	Polymerization in which a tactic polymer is formed. However, polymerization in which stereoisomerism present in the monomer is merely retained in the polymer is not to be regarded as stereospecific. For example, the polymerization of a chiral monomer, e.g. D-propylene oxide (D-methyloxirane), with retention of configuration is not considered to be a stereospecific reaction; however, selective polymerization, with retention, of one of the enantiomers present in a mixture of D- and L-propylene oxide molecules is so classified.
1.12	ditactic polymer	A tactic polymer that contains two sites of defined stereoisomerism in the main chain of the configurational base unit.

Examples

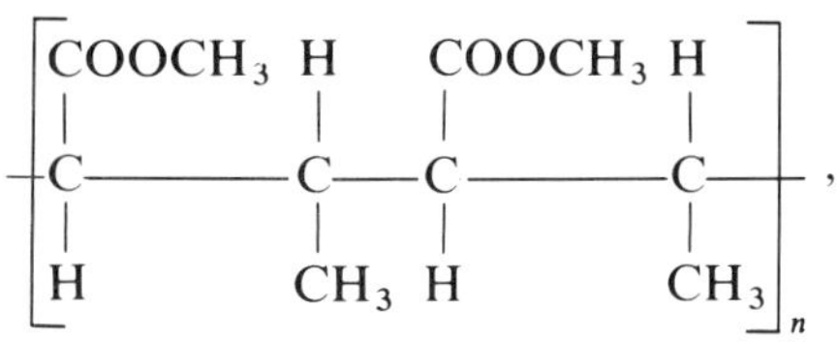,
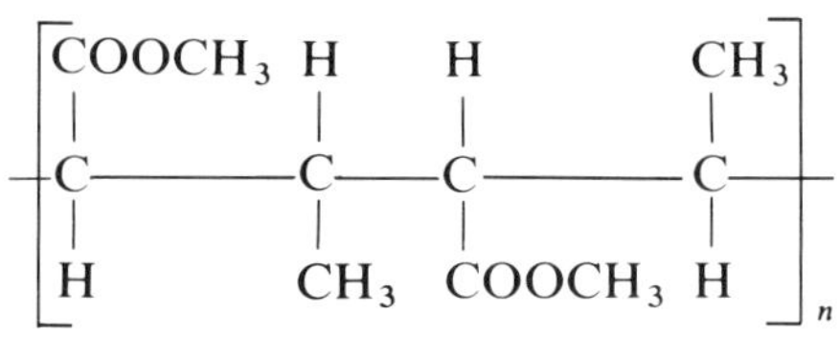

are both ditactic.

Term		*Definition*
1.13	tritactic polymer	A tactic polymer that contains three sites of defined stereoisomerism in the main chain of the configurational base unit.

Example

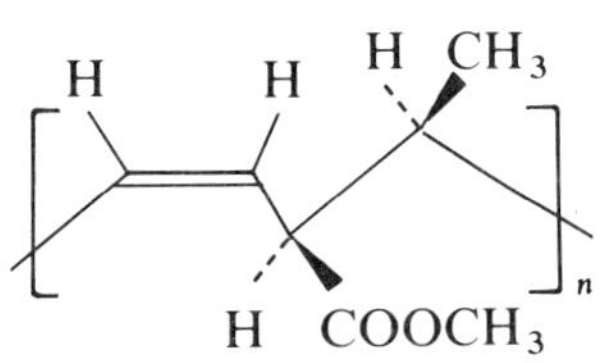

poly[3-(methoxycarbonyl)-4-methyl-*cis*-1-butenylene]

Term		*Definition*
1.14	diisotactic polymer	An isotactic polymer that contains two chiral or prochiral atoms with defined stereochemistry in the main chain of the configurational base unit.
1.15	disyndiotactic polymer	A syndiotactic polymer that contains two chiral or prochiral atoms with defined stereochemistry in the main chain of the configurational base unit.

Examples

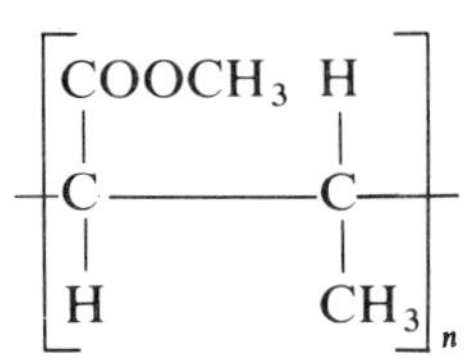,
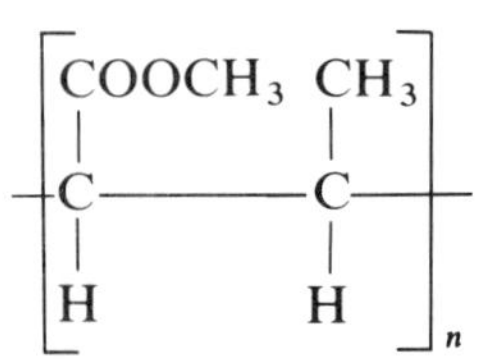

diisotactic

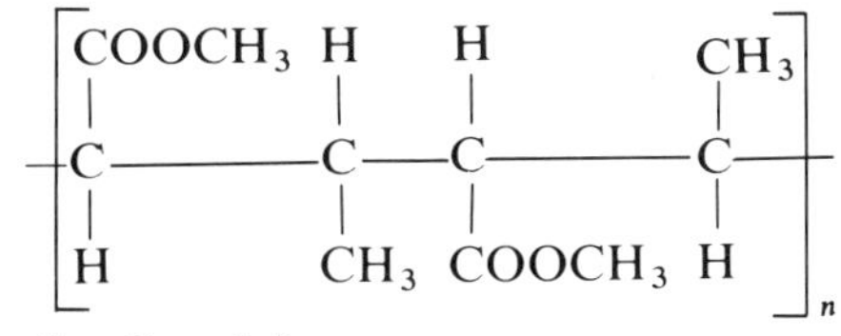

disyndiotactic*

A polymer with the repeating unit

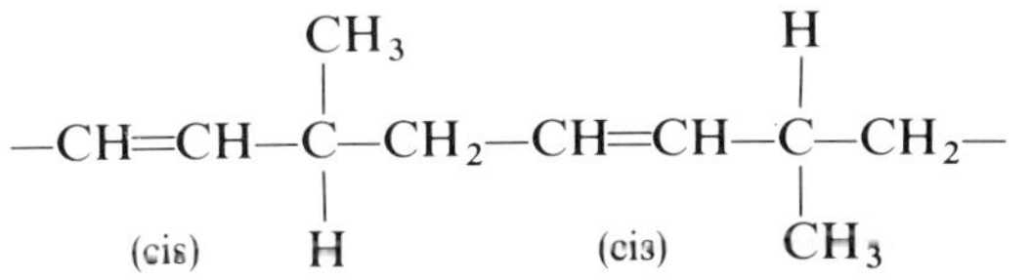

is ditactic and may be described as syndiotactic (see Definition 1.8), but it is *not* disyndiotactic.

The relative configuration of adjacent, constitutionally non-equivalent, carbon atoms can be specified as 'erythro' or 'threo', as appropriate, by adding the required prefix to the terms 'diisotactic' and 'disyndiotactic', as necessary (see Section 2.2).

Term	*Definition*
1.16 cistactic polymer	A tactic polymer in which the main-chain double bonds of the configurational base units are entirely in the cis arrangement.
1.17 transtactic polymer	A tactic polymer in which the main-chain double bonds of the configurational base units are entirely in the trans arrangement.

Terms referring to the tacticity of polymers (tactic, ditactic, tritactic, isotactic, cistactic, etc) can also be applied with similar significance to chains, sequences, blocks, etc.

Note

Terms defining stereochemical arrangements are to be italicized only when they form part of the name of a polymer; the use of such terms as adjectives, even when immediately preceding names, does not require italics. This practice is illustrated in the examples below.

Examples

isotactic poly(3-methyl-*trans*-1-butenylene)

transisotactic poly(3-methyl-1-butenylene)†

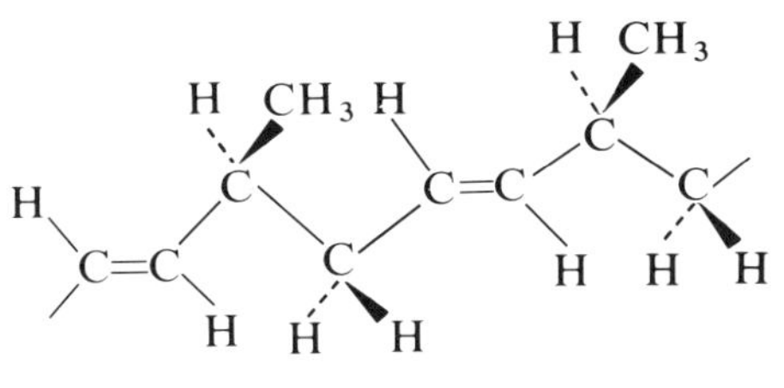

and/or

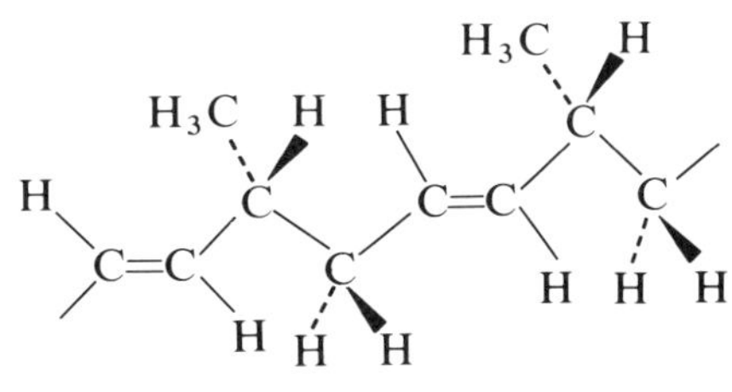

* 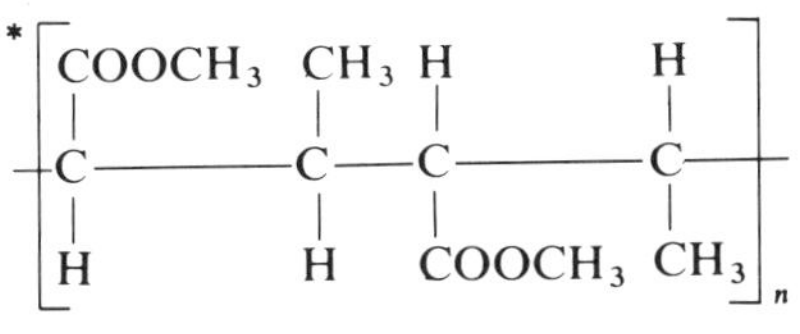 does *not* represent a different disyndiotactic polymer.

†Structure-based name; either may be used.

diisotactic poly[*threo*-3-(methoxycarbonyl)-4-methyl-*trans*-1-butenylene]
transthreodiisotactic poly[3-(methoxycarbonyl)-4-methyl-1-butenylene]*

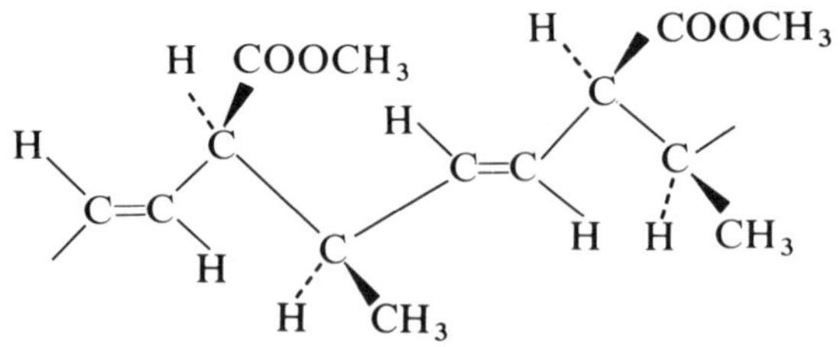

and/or

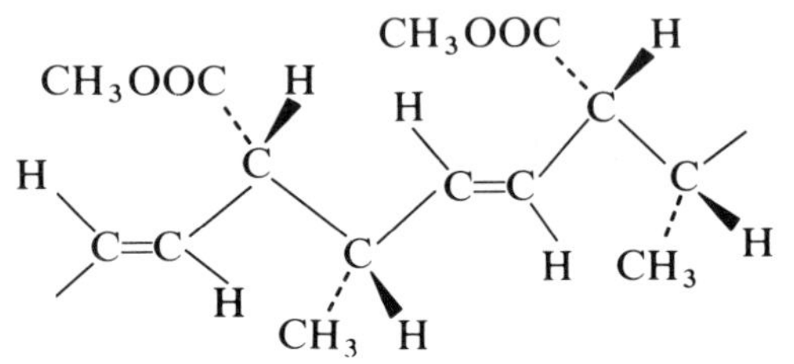

Term	Definition
1.18 block	A portion of a polymer molecule, comprising many constitutional units, that has at least one constitutional or configurational feature which is not present in the adjacent portions.
1.19 tactic block	A regular block that can be described by only one species of configurational repeating unit in a single sequential arrangement.
1.20 atactic block	A regular block that has equal numbers of the possible configurational base units in a random sequence distribution.
1.21 stereoblock	A regular block that can be described by one species of stereorepeating unit in a single sequential arrangement.
1.22 tactic block polymer	A polymer, the molecules of which consist of tactic blocks connected linearly.
1.23 stereoblock polymer	A polymer, the molecules of which consist of stereoblocks connected linearly.

Examples
Tactic block polymer

$—A_k—B_l—A_m—B_n—$

where A and B are, for example,

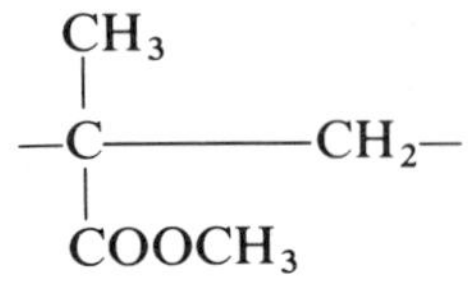

(A)

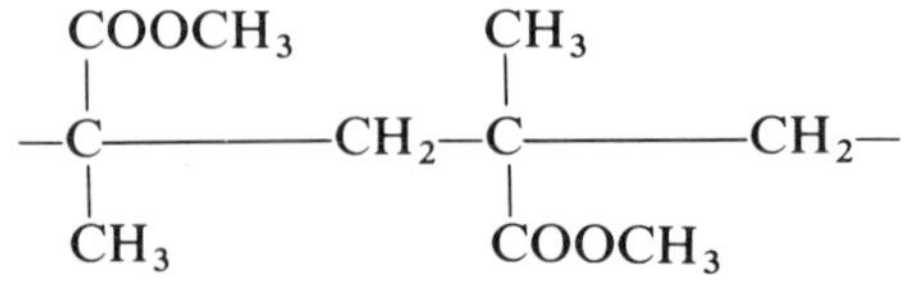

(B)

*Structure-based name; either may be used.

In this case the blocks are stereoblocks but the block polymer is not a block copolymer because all the units derive from a single monomer.

In the following example of a regular poly(propylene) chain, the stereoblocks are denoted by |______| . Here, the sequence of identical relative configurations of adjacent units that characterizes the stereoblock is terminated at each end of the block. Note that ¦_ _ _ _ _¦ represents a configurational sequence, which may or may not be identical with a stereoblock (see Definitions 1.21 and 2.1.3).

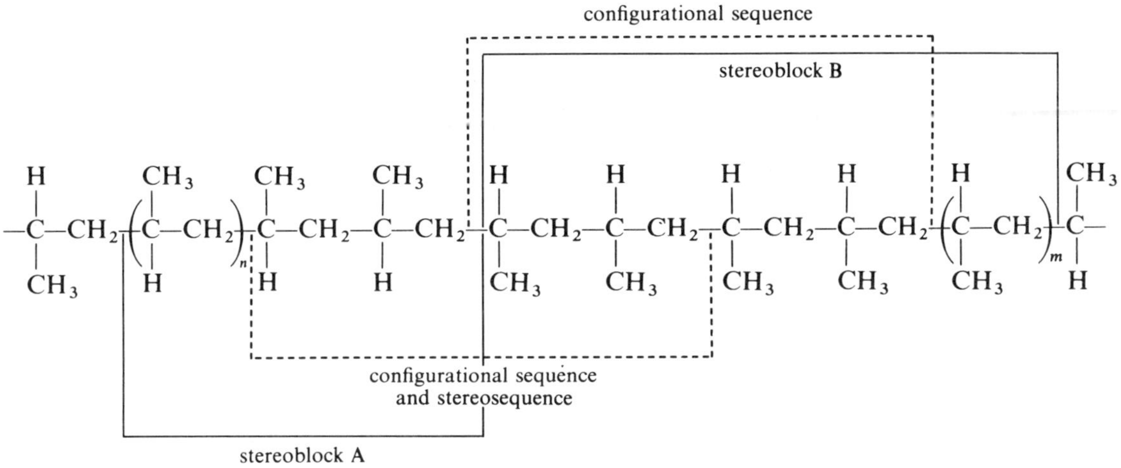

The configurational sequence and stereosequence coincide in this particular case because there is only one site of stereoisomerism in each constitutional repeating unit (compare Definitions 2.1.3 and 2.1.4).

2 SEQUENCES

2.1 Constitutional and configurational sequences

The description of polymer structure revealed by studies of physical properties focuses attention on the distribution of local arrangements present in the molecules, and terms useful in this context are defined below. (The terms defined here in relation to complete polymer molecules can also be applied to sequences and to blocks, as in Reference [2], Definition 3.14.)

Term	*Definition*
2.1.1 constitutional sequence	A defined portion of a polymer molecule comprising constitutional units of one or more species.

Example

$-CH_2-CH_2-CH_2-CH(CH_3)-$ $\quad -CH_2-CH(CH_3)-CH_2-CH(CH_3)-$

Term	*Definition*
2.1.2 constitutional homosequence	A constitutional sequence which contains constitutional units of only one species and in one sequential arrangement.

Example

$-CH(CH_3)-CH_2-CH(CH_3)-CH_2-$ $\quad -\!\!\left[CH(CH_3)-CH_2\right]_6\!\!-$

In these two cases, the constitutional unit —$CH(CH_3)$—CH_2— can be called the constitutional repeating unit of the homosequence.

Term	*Definition*
2.1.3 configurational sequence	A constitutional sequence in which the relative or absolute configuration is defined at one or more sites of stereoisomerism in each constitutional unit in the main chain of a polymer molecule. (See example following Definition 1.23.)
2.1.4 stereosequence	A configurational sequence in which the relative or absolute configuration is defined at all sites of stereoisomerism in the main chain of a polymer molecule.
2.1.5 configurational homosequence	A constitutional homosequence in which the relative or absolute configuration is defined at one or more sites of stereoisomerism in each constitutional unit in the main chain of a polymer molecule.
2.1.6 stereohomosequence	A configurational homosequence in which the relative or absolute configuration is defined at all sites of stereoisomerism in the main chain of a polymer molecule.

2.2 Description of relative configurations

Erythro and threo structures

The relative configuration at two contiguous carbon atoms in the main chain bearing, respectively, substituents a and b ($a \neq b$), is designated by the prefix 'erythro' or 'threo', as appropriate, by analogy with the terminology for carbohydrate systems* in which the substituents are —OH.

Examples

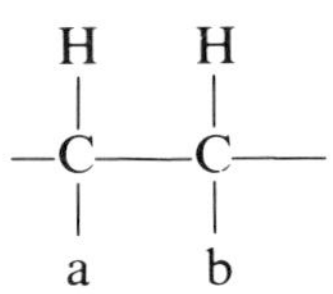

erythro

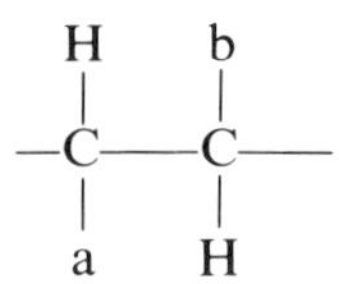

threo

Similar systems in which a higher level of substitution exists may be treated analogously if the erythro or threo designation is employed to denote the relative placements of those two substituents, one for each backbone carbon atom, which rank highest according to the Sequence

*See Rule 4.11 of Reference [4].

Rule. Thus, the following hypothetical examples would be designated as indicated:

Examples

erythro

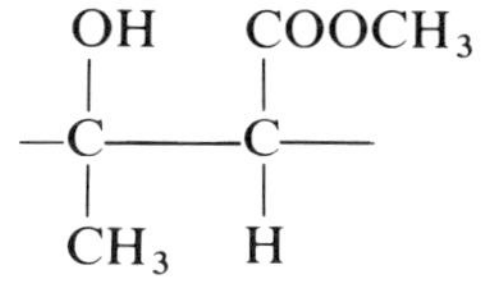

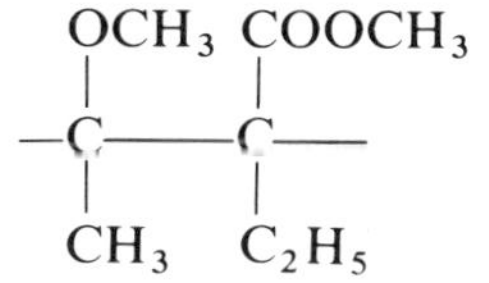

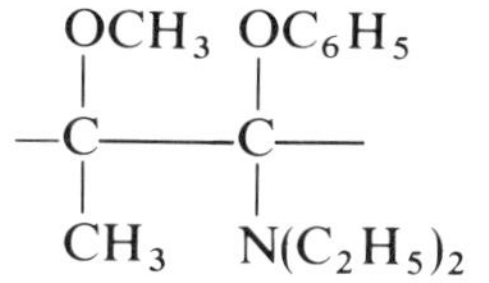

threo

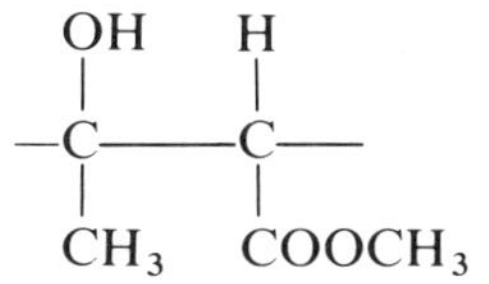

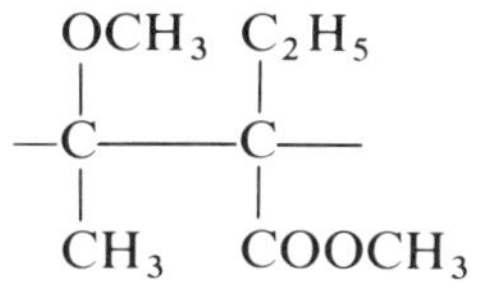

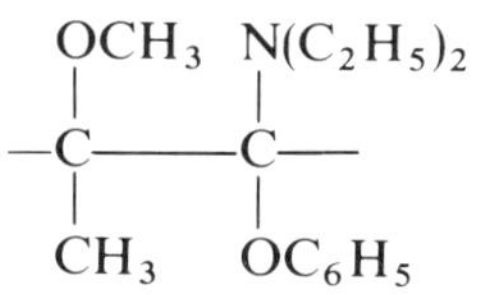

This novel extension of the erythro/threo terminology, especially its conjunction with the Sequence Rule, is specifically proposed solely to cope with the problems incurred in describing the steric structures of macromolecules.

Meso and racemo structures

Relative configurations of consecutive, but not necessarily contiguous, constitutionally equivalent carbon atoms that have a symmetrically constituted connecting group (if any) are designated as 'meso' or 'racemo', as appropriate.

Examples

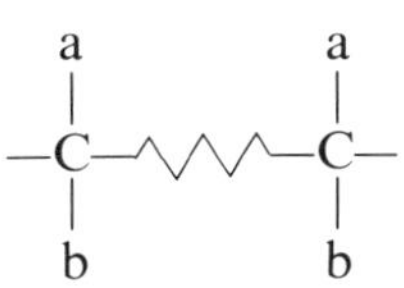 abbreviation *m*

meso

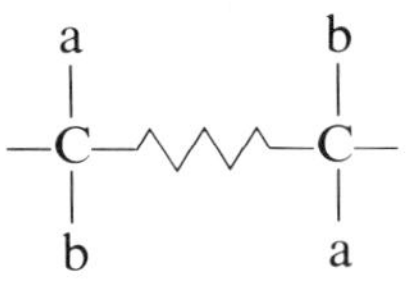
abbreviation *r*

racemo

(The symbol —〰— represents a symmetrically constituted connecting group, such as $-CH_2-$, $-CH_2-CH_2-$, or $-CR_2-CH_2-CR_2-$.)

Note

The structures

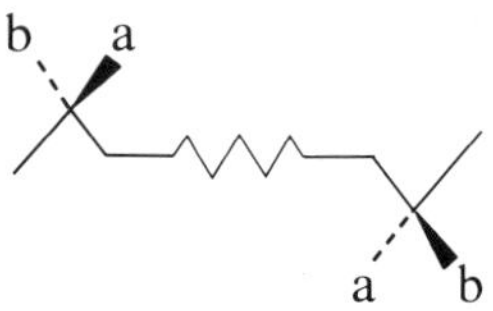

and

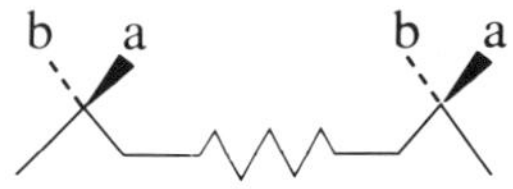

both have the meso relative configuration but the boldly printed carbon atoms in each of the formulae below cannot be considered as in a meso arrangement because the connecting group lacks the necessary symmetry.

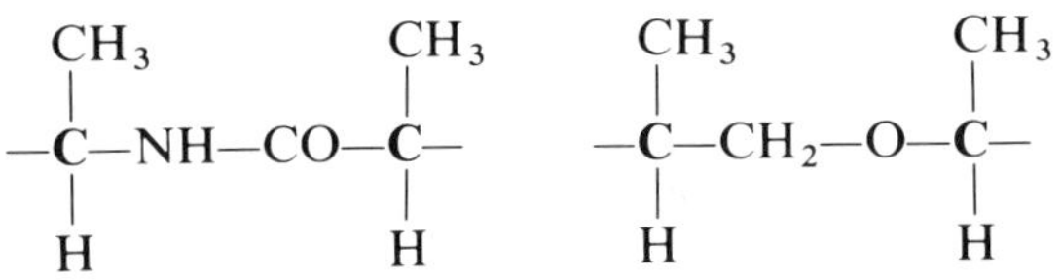

The term 'racemo' is introduced here as the logical prefix for the designation of an arrangement that is analogous to racemic, in the sense defined above. It is unfortunate that the meaning of the term 'racemic' current in organic chemistry is not directly applicable to polymers, but the use of the prefix 'racemo' proposed here should not cause confusion because of the special context. To achieve a full configurational description, it may be necessary to preface the name of a polymer with a compound adjective that combines a term such as 'erythro', 'threo', 'meso' or 'racemo' with a term such as 'diisotactic' or 'disyndiotactic'.

Examples

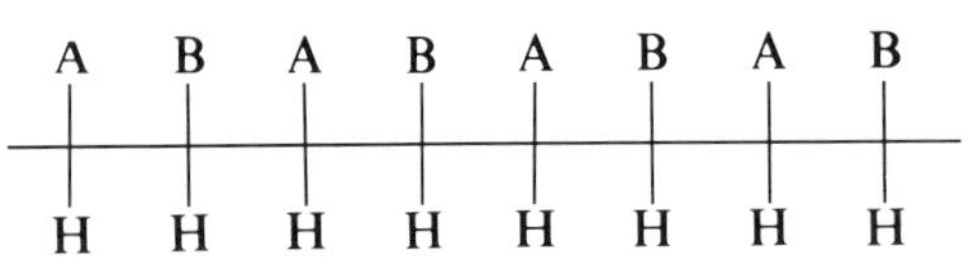

erythrodiisotactic polymer

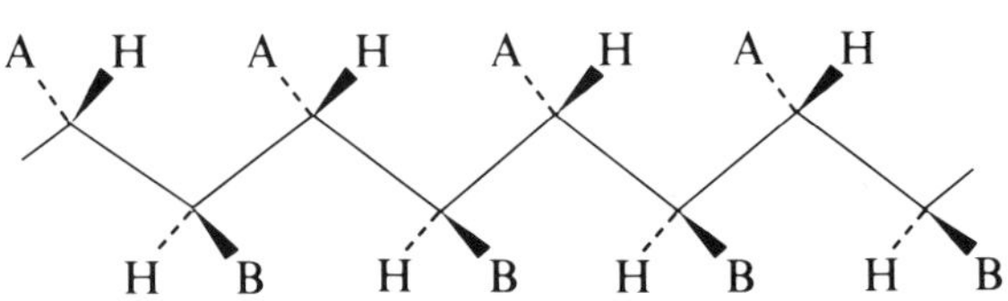

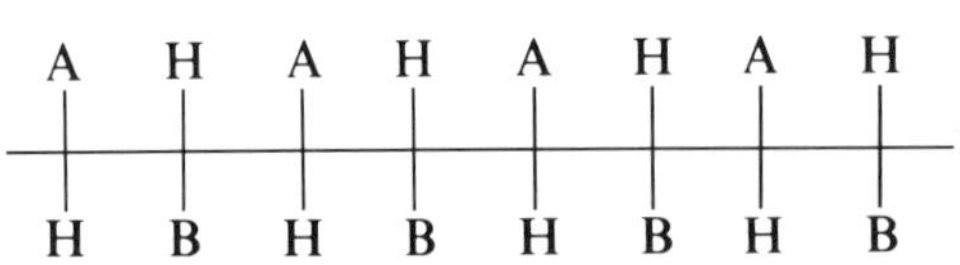

threodiisotactic polymer

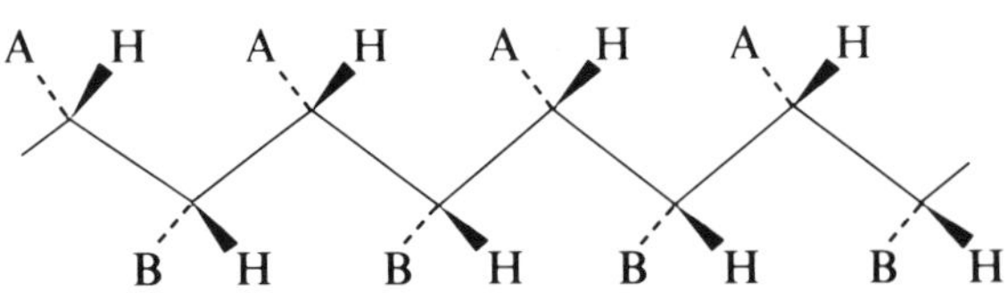

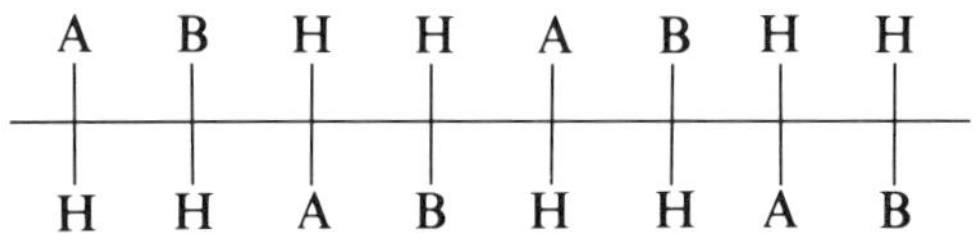

disyndiotactic polymer*

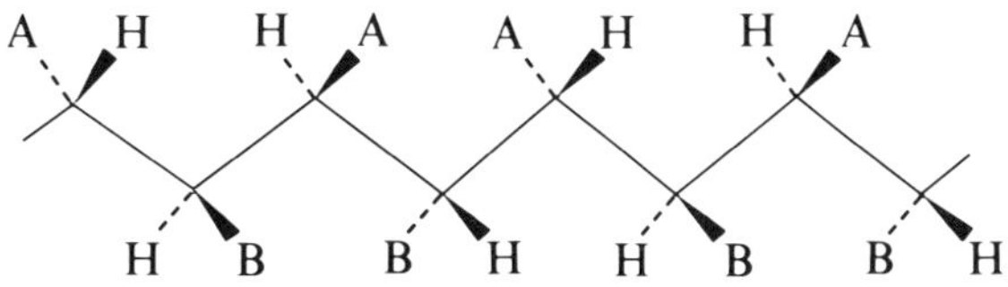

Polymers with chiral centres arising from rings linking adjacent main-chain carbon atoms can be included in this nomenclature:

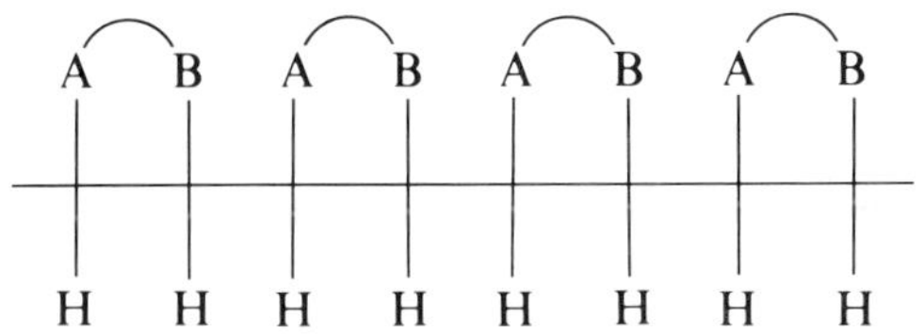

erythrodiisotactic polymer

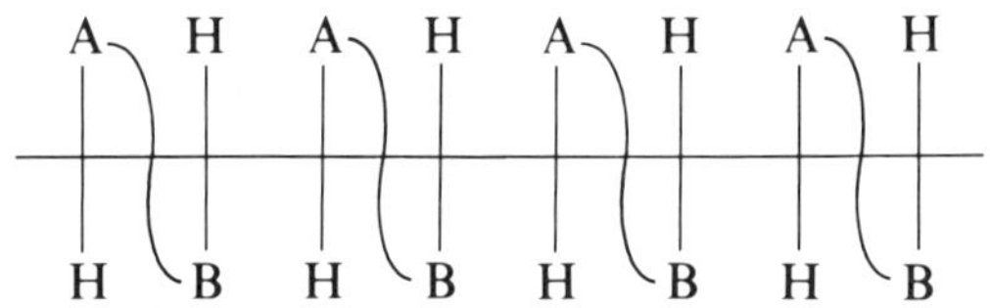

threodiisotactic polymer

In the last two cases, the chiralities of the asymmetric centres should be designated *R*- or *S*-, if known.

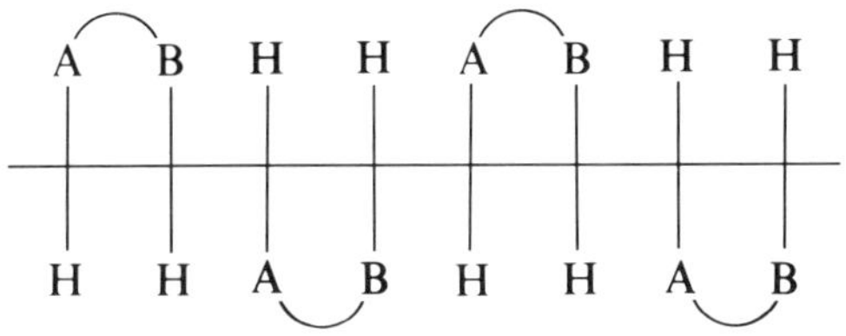

erythrodisyndiotactic polymer

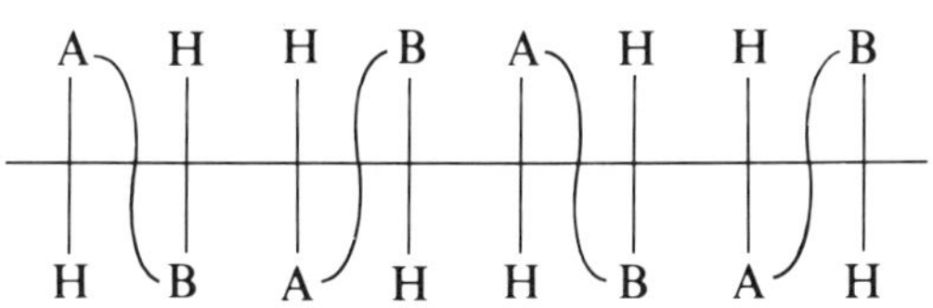

threodisyndiotactic polymer

*This polymer cannot be expressed as erythrodisyndiotactic nor as threodisyndiotactic. Instead:

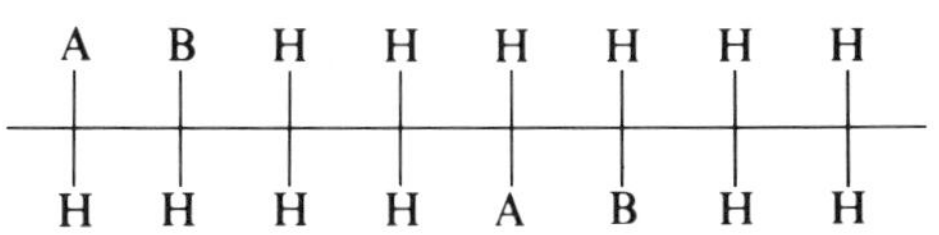

erythrodisyndiotactic

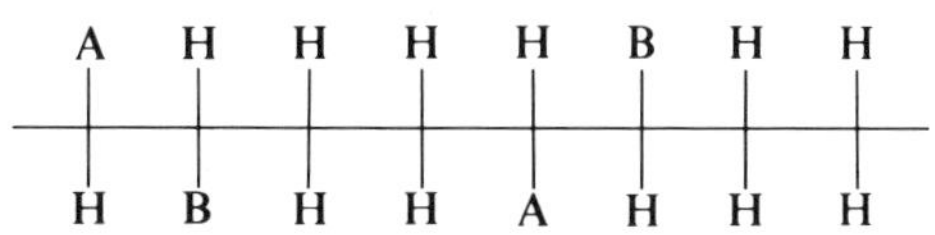

threodisyndiotactic

If the rings are symmetrical:

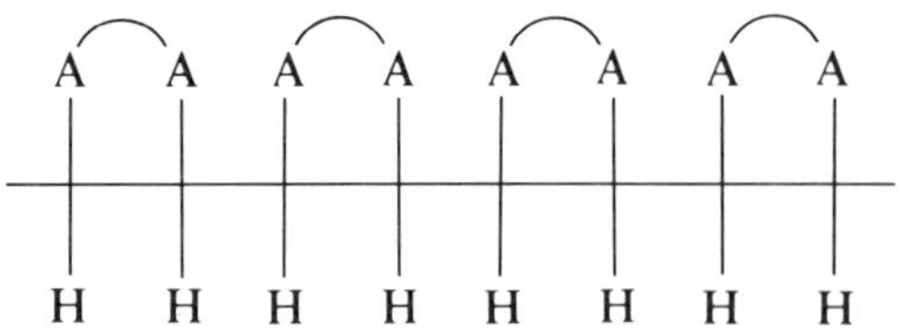

mesodiisotactic

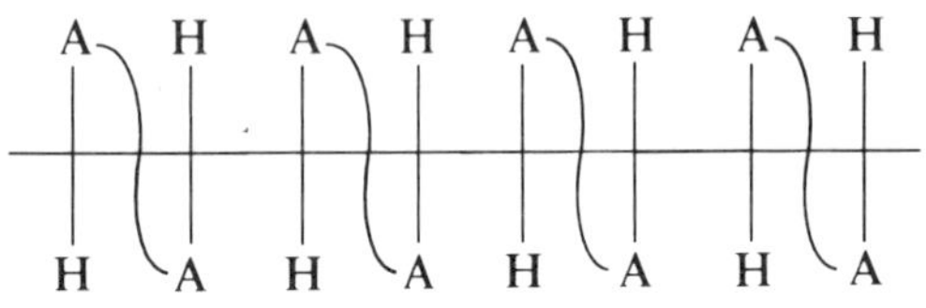

racemodiisotactic

Stereosequences
Stereosequences terminating in tetrahedral stereoisomeric centres at both ends, and which comprise two, three, four, five, etc consecutive centres of that type, may be called diads, triads, tetrads, pentads, etc, respectively.

Typical diads are:

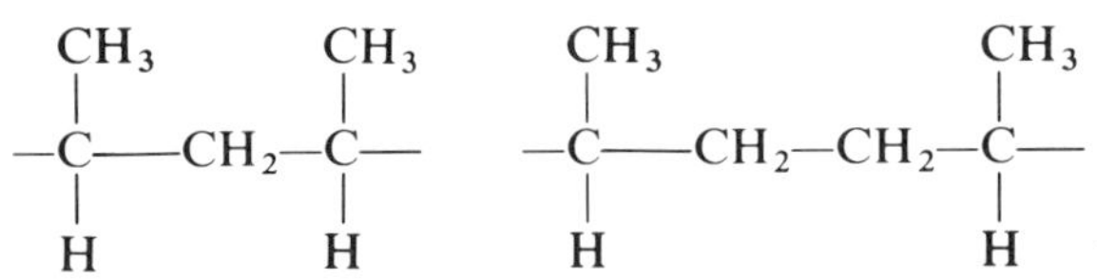

When it is necessary to specify the internal stereochemistry of the group, a prefix is required. In vinyl polymers there are meso (*m*) and racemic (*r*) diads and *mm*, *mr*, *rr* triads. The latter may be called isotactic, heterotactic and syndiotactic triads, respectively. Stereoregular vinyl polymers can be defined in terms of the regular sequences of diads; thus an isotactic vinyl polymer consists entirely of *m* diads, i.e. it corresponds to the following succession of relative configuration —*m m m m m m*—, whereas a syndiotactic vinyl polymer consists entirely of *r* diads, corresponding to the sequence —*r r r r r r*—. Similarly, a vinyl polymer consisting entirely of *mr* (=*rm*) triads is called a heterotactic polymer.

3 CONFORMATIONS

3.1 Designation of conformation of polymer molecules

Bond lengths
If a specific A—B bond is denoted as A_i—B_j, the bond length is written $b(A_i, B_j)$. Abbreviated notations, such as b_i, may be used if this meaning is clarified by a diagram.

Bond angles

The bond angle formed by three consecutive atoms

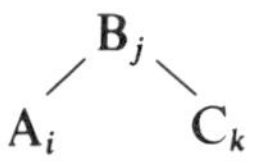

is written $\tau(A_i, B_j, C_k)$ which may be abbreviated, if there is no ambiguity, to $\tau(B_j)$, τ_j^B, $\tau B(j)$ or τ_j.

Torsion angles

If a system of four consecutive atoms

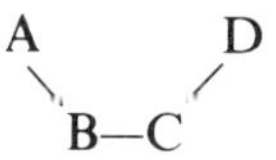

is projected onto a plane normal to bond B—C, the angle between the projection of A—B and the projection of C—D is described as the torsion angle of A and D about bond B—C; this angle may also be described as the angle between the plane containing A, B and C and the plane containing B, C and D. The torsion angle is written in full as $\theta(A_i, B_j, C_k, D_l)$, which may be abbreviated, if there is no ambiguity, to $\theta(B_j, C_k)$, $\theta(B_j)$ or θ_j^B, etc. In the eclipsed conformation in which the projections of A—B and C—D coincide, θ is given the value 0° (synperiplanar conformation). A torsion angle is considered positive $(+\theta)$ or negative $(-\theta)$ depending on whether, when the system is viewed along the central bond in the direction B——C (or C——B), the bond to the front atom A (or D) requires the smaller angle of rotation to the right or to the left, respectively, in order that it may eclipse the bond to the rear atom D (or A); note that it is immaterial whether the system be viewed from one end or the other. According to this definition, a sequence of consecutive positive torsion angles generates a right-handed helix (Helix sense, p. 43). It is to be noted that:

1. torsion angles are measured in the range $-180° < \theta < +180°$ rather than from 0° to 360°, so that the relationship between enantiomeric configurations or conformations can be readily appreciated;
2. any Greek letter from the end of the alphabet, except τ, can be used to denote torsion angles; θ or ω are recommended;
3. abbreviated notations are preferably restricted to bond lengths, bond angles and torsion angles related to main-chain atoms.

Conformations referring to torsion angles $\theta(A, B, C, D)$, where A, B, C, D are main-chain atoms, can be described as: *cis* or *synperiplanar* (C); *gauche* or *synclinal* (G); *anticlinal* (A); and *trans* or *antiperiplanar* (T), corresponding to torsion angles within $\pm 30°$ of, respectively, 0°, $\pm 60°$, $\pm 120°$ and $\pm 180°$. The letters shown in parentheses (upper case C, G, A, T) are the recommended abbreviations.*

The symbols G^+, G^- (or A^+, A^-, for example) refer to torsion angles of similar type but opposite known sign, i.e. $\sim +60°$, $\sim -60°$ (or $\sim +120°$, $\sim -120°$). The notation G, $\bar{G}$; A, $\bar{A}$ (and T, $\bar{T}$; C, $\bar{C}$ whenever the torsion angles are not exactly equal to 180° and 0°, respectively) is reserved for the designation of enantiomorph conformations, i.e. conformations of opposite but unspecified sign. Where necessary, a deviation from the proper value of the torsion angle can be indicated by the sign $(\sim)$, as in the following examples: $G(\sim)$; $\bar{G}(\sim)$; $G^+(\sim)$; $G^-(\sim)$.

*Different authors variously use upper and lower case letters in this context. The desire for uniformity necessitates an arbitrary choice between the alternatives, and the upper case has been selected in the belief that it conflicts less with other designations, for example, the use of *c* and *t* on pages 42 and 43.

Examples

The chain conformation of isotactic poly(propylene) in the crystalline state is:

...TGTGTGTG...

The chain conformation of syndiotactic poly(propylene) in the crystalline state is:

...TTGGTTGG... or ...TTTTTTTT...

The chain conformation of a right-handed α-helix is:

...G^-G^-(trans) G^-G^-(trans)...

or ...G^-(~)G^-(~)(trans)G^-(~)G^-(~)(trans)...

The symbols (cis) and (trans) may be used to designate rigid dihedral angles such as those occurring with double bonds.

The chain conformation of crystalline poly(1, 1-difluoroethylene), modification 2, is:

...$TGT\bar{G}TGT\bar{G}$...

The chain conformation of crystalline poly(*trans*-1-methyl-1-butenylene) in the α-form is:

...(trans) $CT\bar{A}$(trans) $CT\bar{A}$...

The chain conformation of isotactic vinyl polymers in the crystalline state is:

...T(~)G(~)T(~)G(~)...

A possible conformation of isotactic poly(propylene) in the melt can be described as:

$$\ldots(\mathrm{TG})_{n_1}(\mathrm{TT})(\bar{\mathrm{G}}\mathrm{T})_{m_1} \begin{matrix} (\bar{\mathrm{A}}\mathrm{G}) \\ \text{or} \\ (\bar{\mathrm{G}}\mathrm{A}) \end{matrix} (\mathrm{TG})_{n_2}(\mathrm{TT})(\bar{\mathrm{G}}\mathrm{T})_{m_2} \begin{matrix} (\bar{\mathrm{A}}\mathrm{G}) \\ \text{or} \\ (\bar{\mathrm{G}}\mathrm{A}) \end{matrix} (\mathrm{TG})_{n_3}$$

3.2 Specific terminology for crystalline polymers

The crystallographic identity period parallel to the chain axis should preferably be designated c in descriptions of macromolecular crystallography.

In the description of helices, the following parameters and symbols should be employed: n signifies the number of conformational repeating units per turn (the conformational repeating unit in a crystalline polymer is the smallest unit of given conformation that is repeated through symmetry operations which comprise a translation. In most cases it corresponds to the configurational repeating unit); h signifies the unit height, i.e. the translation along the helix axis per conformational repeating unit; t signifies the unit twist, i.e. the angle of rotation about the helix axis per conformational repeating unit.

Examples

If the number of conformational repeating units along the identity period c is M and the number of turns is N, then:

$n = M/N$

$h = c/M$

$t = 2\pi N/M$

For isotactic poly(propylene), since $M=3$, $N=1$ and $c=6.50$ Å,

$n=3$
$h=2.17$ Å
$t=2\pi/3$

For poly(oxymethylene), since $M=9$, $N=5$ and $c=17.39$ Å,

$n=1.8$
$h=1.93$ Å
$t=2\pi(5/9)$.

Helix sense
The right-handed sense of a helix traces out a clockwise rotation moving away from the observer; the left-handed sense of a helix traces out an anticlockwise rotation moving away from the observer, e.g. the ...$TG^+TG^+TG^+$... helix of isotactic poly(propylene) is left-handed.

Isomorphous and enantiomorphous structures
In the crystalline state, polymer chains are generally parallel to one another but neighbouring chains of equivalent conformation may differ in chirality and/or orientation.

Chains of identical chirality and conformation are *isomorphous.* Chains of opposite chirality but equivalent conformation are *enantiomorphous.*

Examples
Two ...$TG^+TG^+TG^+$... helices of isotactic poly(propylene) are isomorphous.

Isotactic poly(propylene) chains of the ...$TG^+TG^+TG^+$... and ...$G^-TG^-TG^-T$... types are mutually enantiomorphous.

Isoclined and anticlined structures
With regard to orientation, consider a repeating side group originating at atom A_1^i, the first atom of the side group being B_α^i. For certain chain symmetries (helical, for instance) the bond vectors

$\vec{\mathbf{b}}(A_1^i, B_\alpha^i)$

have the same components (positive or negative)

$\vec{\mathbf{b}}\cdot\vec{\mathbf{c}}/|\vec{\mathbf{c}}|$

along the c axis for every i.

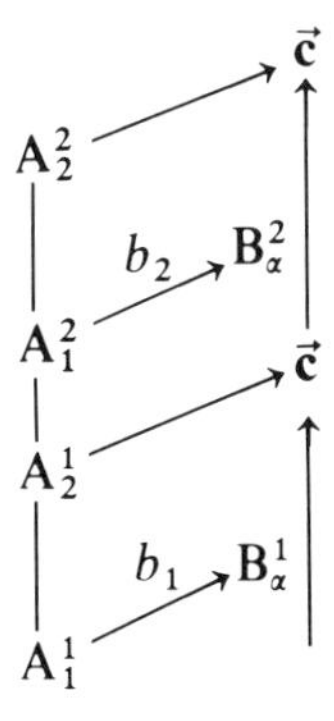

Two equivalent (isomorphous or enantiomorphous) chains in the crystal lattice, having identical components of the bond vectors along c, both positive or both negative, are designated *isoclined*; two equivalent chains having bond vectors along *c* of the same magnitude but opposite sign are designated *anticlined.*

Examples

1. Isotactic poly(3-methyl-*cis*-1-butenylene)

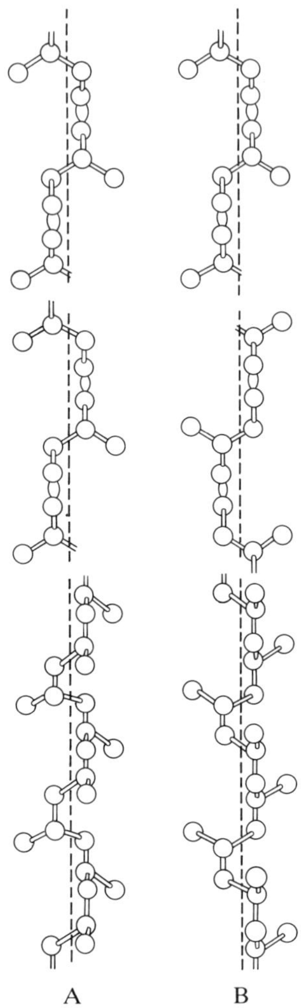

Isoclined isomorphous chains: the two chains have parallel axes and the same orientation of the pendant methyl groups.

Anticlined isomorphous chains: the two chains have parallel axes and opposite orientation of the pendant methyl groups.

2. Isotactic poly(propylene)

Anticlined enantiomorphous chains: the conformation of A corresponds to a $(TG^-)_n$ bond succession (right-handed helix). The conformation of B corresponds to a $(G^+T)_n$ bond succession (left-handed helix).

Line repetition groups and symmetry elements

To designate linear chain conformations in the crystalline state, the use of line repetition groups [6] is recommended.

First symbol t translation
s screw repetition

[In this case of screw repetition, the number of conformational repeating residues per turn is included in parentheses, i.e.: s(11/3) or s(3.67$\pm$0.02).]

Second and further symbols. The symmetry elements required to define the line repetition group are suggested in Reference [7]. Possible symmetry elements are:

i centre of symmetry
m plane of symmetry perpendicular to the chain axis

c glide plane parallel to the chain axis
d plane of symmetry parallel to the chain axis
2 two-fold axis of symmetry perpendicular to the chain axis

The possible line repetition groups are listed below, with examples. (The structure-based name is given first, the process-based name second, in each case.)

Table 2.1 Chain symmetry of some crystalline polymers

Line repetition group	Polymer
t1	poly(1-methyl-*trans*-1-butenylene) trans-1,4-polyisoprene
s(*M*/*N*)1	isotactic poly(propylene) ($M/N=3/1$) isotactic polypropene ($M/N=3/1$)
s(*M*/*N*)2	syndiotactic poly(propylene) ($M/N=2/1$) (helical modification) syndiotactic polypropene ($M/N=2/1$) (helical modification)
t*m*	poly[imino(1,7-dioxoheptamethylene)iminoheptamethylene] poly(heptamethylene pimelamide)
t*c*	poly(1,1-difluoroethylene) (modification 2) poly(1,1-difluoroethylene) (modification 2)
t*i*	diisotactic poly(1,2-dimethyltetramethylene) diisotactic alternating copolymer of ethylene and *cis*-2-butene
s(2/1)*m*	poly(1-pentenylene) poly(cyclopentene)
s(2/1)*d*	poly(iminoadipoyliminohexamethylene) poly(hexamethylene adipamide)
t*dm*	—
t*id*	—
t*cm*	syndiotactic poly(1-vinylethylene) syndiotactic 1,2-poly(1,3-butadiene)
s(2/1)*dm*	poly(methylene) polyethylene

4 SUPPLEMENTARY DEFINITIONS*

Term	*Definition*
4.1 degrees of triad isotacticity, syndiotacticity and heterotacticity	The fractions of triads in a regular vinyl polymer that are of the *mm*, *rr* and *mr* = *rm* types, respectively. In cases where triad analysis is not attainable, the diad isotacticity and diad syndiotacticity may be defined as the fractions of diads in a regular-vinyl polymer that are of the *m* and *r* types, respectively.

*Tacticity in side chains will be treated in a later document.

Term	*Definition*
4.2 degrees of cistacticity and transtacticity	For a regular polymer containing double bonds in the main chain of the constitutional repeating units, these are the fractions of such double bonds that are in the cis and trans configurations, respectively.
4.3 degree of crystallinity	This is defined as the fraction of the bulk polymer that consists of regions showing long-range three-dimensional order.
4.4 lateral order	Order in the side-by-side packing of the molecules of a linear polymer.
4.5 longitudinal order	Order in the atomic positions along the chains of a linear polymer.

Acknowledgements
Members of the Commission who have contributed actively to the preparation of the document are: Prof. G. Allegra, Dr N.M. Bikales (Secretary 1979 – 1987), Prof. P. Corradini, Dr L.C. Cross, Dr R.B. Fox (Secretary 1973–1979), Prof. A.D. Jenkins (Chairman 1977–1985), Dr J.F. Kennedy, Dr P. Kratochvíl, Dr K.L. Loening (Chairman 1968–1977), Prof. I.M. Papisov, Prof. N.A. Platé, Prof. W. Ring, Prof. P. Sigwalt, Dr U.W. Suter, Prof. T. Tsuruta and Dr R.E. Wetton.

REFERENCES

1 M.L. Huggins, G. Natta, V. Desreux and H. Mark (for IUPAC Commission on Macromolecules). Report on nomenclature dealing with steric regularity in high polymers. *J. Polym. Sci.* **56**, 153–161 (1962); *Pure Appl. Chem.* **12**, 643–656 (1966).
2 IUPAC. Basic definitions of terms relating to polymers 1974. *Pure Appl. Chem.* **40**, 477–491 (1974). Reprinted as Chapter 1, this volume.
3 IUPAC. Nomenclature of regular single-strand organic polymers (rules approved 1975). *Pure Appl. Chem.* **48**, 373–385 (1976). Reprinted as Chapter 5, this volume.
4 IUPAC. Rules for the nomenclature of organic chemistry. Section E: stereochemistry (recommendations 1974). *Pure Appl. Chem.* **45**, 11 (1976).
5 IUPAC-IUB. Abbreviations and symbols for the description of the conformation of polypeptide chains (rules approved 1974). *Pure Appl. Chem.* **40**, 291–308 (1974).
6 A. Klug, F.AC Crick and H.W. Wyckoff. *Acta Crystallogr.* **11**, 199 (1958).
7 P. Corradini. Chain conformation and crystallinity. In *Stereochemistry of Macromolecules* (Ed. A. Ketley). Marcel Dekker, New York, Part III, 1–60 (1968).

3: Definitions of Terms Relating to Individual Macromolecules, Their Assemblies, and Dilute Polymer Solutions (1988)

PREAMBLE

This document is part of a series published by the Commission on Macromolecular Nomenclature dealing with definitions for the important terms in polymer science [1–3]. It is also the first of several documents that will be published dealing specifically with physicochemical terminology in the polymer field. The recommendations presented here deal with such key areas of the physical chemistry of macromolecules as individual macromolecules, their assemblies and dilute polymer solutions; they include recommended terminology for molecular weight, molecular-weight averages, distribution functions, radius of gyration, the Flory–Huggins theory, viscosity of solutions, scattering of radiation by polymers, fractionation, etc.

The reader's attention is especially directed to the new terms 'uniform polymer' and 'non-uniform polymer' which denote polymers composed of molecules that are uniform or non-uniform, respectively, with respect to relative molecular mass and constitution. These terms replaced the widely used, but non-descriptive and self-contradictory terms 'monodisperse polymer' and 'polydisperse polymer'.

1 INDIVIDUAL MACROMOLECULES

Term	*Definition*
1.1 relative molecular mass, molecular weight	recommended symbol: M_r The ratio of the average mass per formula unit of a substance to 1/12 of the mass of an atom of nuclide ^{12}C. *Note* See Definition 1.2.
1.2 molar mass	recommended symbol: M Mass divided by amount of substance. *Notes* 1. Molar mass is usually expressed in g/mol or kg/mol units. The g/mol unit is recommended in polymer science, since then the numerical values of the molar mass and the relative molecular mass of a substance are equal.

Prepared by a working group consisting of P. Kratochvíl (Czechoslovakia) and U.W. Suter (Switzerland). Reprinted from *Pure Appl. Chem.* **61**, 211–241 (1989).

	Term	*Definition*
		2. Relative molecular mass (molecular weight) is a pure number and must not be associated with any units. 3. The use of the dalton as a unit of mass, identical with the atomic mass unit, is discouraged. 4. The terms 'molar' and 'molecular' may also be used for particles consisting of more than one molecule, such as complexes, aggregates, micelles, etc. 5. If there is no danger of confusion, the subscript r in the recommended symbol for the relative molecular mass, M_r, may be omitted.
1.3	degree of polymerization	recommended symbol: X recommended abbreviation: DP The number of monomeric units in a molecule of a polymer.
1.4	thermodynamically equivalent sphere	A sphere, impenetrable to other spheres, displaying the same excluded volume (see Definition 3.1.9) as an actual polymer molecule.
1.5	short-range intramolecular interaction	Steric or other interaction involving atoms or groups or both situated nearby in sequence along the chain. *Notes* 1. The interacting atoms or groups are typically separated by fewer than ten consecutive bonds in a chain. 2. If no confusion can occur, the word 'intramolecular' may be omitted.
1.6	long-range intramolecular interaction	Interaction between segments, widely separated in sequence along the chain, that occasionally approach one another during molecular flexing. *Notes* 1. This type of interaction is closely related to the excluded volume of a segment (see Definition 3.1.8), the latter quantity reflecting interactions involving segments and solvent molecules. 2. If no confusion can occur, the word 'intramolecular' may be omitted.
1.7	unperturbed dimensions	The dimensions of an actual polymer random coil (see Definition 1.15) in a theta state (see Definition 3.1.5).
1.8	perturbed dimensions	The dimensions of an actual polymer random coil (see Definition 1.15) not in a theta state (see Definition 3.1.5).
1.9	radius of gyration	recommended symbols: s, $\langle s^2 \rangle^{1/2}$ A parameter characterizing the size of a particle of any shape. For a rigid particle consisting of mass elements of mass m_i,

Term	Definition
	each located at a distance r_i from the centre of mass, the radius of gyration, s, is defined as the square root of the mass-average of r_i^2 for all the mass elements, i.e.
	$$s=\left(\sum_i m_i r_i^2 \Big/ \sum_i m_i\right)^{1/2}$$
	For a non-rigid particle, an average over all conformations is considered, i.e.
	$$\langle s^2\rangle^{1/2}=\left\langle\sum_i m_i r_i^2\right\rangle^{1/2} \Big/ \left(\sum_i m_i\right)^{1/2}$$
	Note The subscript zero is used to indicate unperturbed dimensions (see Definition 1.7), as in $\langle s^2\rangle_0^{1/2}$.
1.10 end-to-end vector	recommended symbol: $\boldsymbol{r}$ The vector connecting the two ends of a linear polymer chain in a particular conformation.
1.11 end-to-end distance	recommended symbol: r The length of the end-to-end vector (see Definition 1.10).
1.12 root-mean-square end-to-end distance	recommended symbol: $\langle r^2\rangle^{1/2}$ The square root of the mean-square end-to-end distance (see Definition 1.11) of a linear polymer chain averaged over all conformations of the chain. For a freely jointed chain (see Definition 1.16) consisting of N segments each of length L, $\langle r^2\rangle^{1/2}=N^{1/2}L$.
	Notes 1. The subscript zero is used to indicate unperturbed dimensions (see Definition 1.7), as in $\langle r^2\rangle_0^{1/2}$. 2. If this term is used repeatedly, and if it is not confusing, the abbreviated name 'end-to-end distance' may be used.
1.13 characteristic ratio	recommended symbol: C_N (C_∞ when $N\to\infty$) The ratio of the mean-square end-to-end distance (see Definition 1.12), $\langle r^2\rangle_0$, of a linear polymer chain in a theta state (see Definition 3.1.5) to $N\cdot L^2$, where N is the number of rigid sections in the main chain, each of length L; if all of the rigid sections are not of equal length, the mean-square value of L is used, i.e.
	$$L^2=\sum_i \overline{L_i^2}/N$$
	Note In simple single-strand chains, the bonds are taken as the rigid sections.

Term		*Definition*
1.14	contour length	The maximum end-to-end distance (see Definition 1.11) of a linear polymer chain. *Notes* 1. For a single-strand polymer molecule, this usually means the end-to-end distance of the chain extended to the all-trans conformation. For chains with complex structure, only an approximate value of the contour length may be accessible. 2. The sum of the lengths of all skeletal bonds of a single-strand polymer molecule is occasionally termed 'contour length'. This use of the term is discouraged.
1.15	random coil, statistical coil	The complete set of spatial arrangements of a chain molecule with a large number of segments that randomly change mutual orientation with time, under conditions in which it is free from external constraints that would affect its conformation. *Note* If the solution of the chain molecules is not in a theta state (see Definition 3.1.5), the segments change mutual orientation only approximately randomly.
1.16	freely jointed chain	A hypothetical linear chain molecule consisting of infinitely thin rectilinear segments uniform in length; each segment can take all orientations in space with equal probability, independently of its neighbours. *Note* For models in which the segments are not all uniform in length, the name 'random-walk chain' has been used.
1.17	equivalent chain	A hypothetical freely jointed chain (see Definition 1.16) with the same mean-square end-to-end distance (see Definition 1.11) and contour length (see Definition 1.14) as an actual polymer chain in a theta state (see Definition 3.1.5).
1.18	statistical segment	A segment of an actual polymer chain which behaves, with respect to some property, virtually as a segment of a freely jointed chain (see Definition 1.16).
1.19	freely rotating chain	A hypothetical linear chain molecule, free from short-range and long-range interactions (see Definitions 1.5 and 1.6), consisting of infinitely thin rectilinear segments (bonds) of fixed length, jointed at fixed bond angles; the torsion angles of the bonds can assume all values with equal probability.
1.20	steric factor	recommended symbol: σ The ratio of the root-mean-square end-to-end distance (see

Term	*Definition*
	Definition 1.12) of a polymer chain with unperturbed dimensions (see Definition 1.7), $\langle r^2 \rangle_0^{1/2}$, to that of a freely rotating chain (see Definition 1.19) with the same structure, $\langle r^2 \rangle_{0,f}^{1/2}$, i.e., $(\langle r^2 \rangle_0 / \langle r^2 \rangle_{0,f})^{1/2}$, in the limit of infinite chain length. *Note* The steric factor reflects the effect of hindrance to free rotation.
1.21 worm-like chain, continuously curved chain	A hypothetical linear macromolecule consisting of an infinitely thin chain of continuous curvature; the direction of curvature at any point is random. *Notes* 1. The model describes the whole spectrum of chains with different degrees of chain stiffness from rigid rods to random coils (see Definition 1.15), and is particularly useful for representing stiff chains. 2. In the literature this chain is sometimes referred to as Porod–Kratky chain.
1.22 persistence length	recommended symbol: a The average projection of the end-to-end vector (see Definition 1.10) on the tangent to the chain contour at a chain end in the limit of infinite chain length. *Note* The persistence length is the basic characteristic of the worm-like chain (see Definition 1.21).
1.23 short-chain branch	An oligomeric offshoot from a macromolecular chain.
1.24 long-chain branch	A polymeric offshoot from a macromolecular chain.
1.25 branching index	recommended symbol: g A parameter characterizing the effect of long-chain branches (see Definition 1.24) on the size of a branched macromolecule in solution and defined as the ratio of the mean-square radius of gyration (see Definition 1.9) of a branched molecule, $\langle s_b^2 \rangle$, to that of an otherwise identical linear molecule, $\langle s_l^2 \rangle$, with the same relative molecular mass (see Definition 1.1) in the same solvent and at the same temperature, i.e. $g = \langle s_b^2 \rangle / \langle s_l^2 \rangle$.
1.26 network	A highly ramified polymer structure in which each constitutional unit is connected to each other constitutional unit and to the macroscopic phase boundary by many permanent paths through the structure, their number increasing with the average number

Term	Definition
	of intervening bonds; these paths must on the average be coextensive with this structure. *Note* Usually, and in all systems that exhibit rubber elasticity, the number of distinct paths is very high, but some constitutional units exist, in most cases, which are connected by a single path only. Sometimes, a structure without any multiple path has also been called a network.
1.27 microgel	A network of microscopic dimensions.
1.28 copolymer micelle	A micelle formed by one or more block or graft copolymer molecules in a selective solvent (see Definition 3.1.14).

2 ASSEMBLIES OF MACROMOLECULES

Term	Definition
2.1 compositional heterogeneity	The variation in elemental composition from molecule to molecule usually found in copolymers.
2.2 constitutional heterogeneity	The variation in constitution from molecule to molecule in polymers with molecules uniform with respect to elemental composition. *Note* An example is a polymer composed of linear and branched molecules; another example is a statistical copolymer comprising two isomeric constitutional units.
2.3 uniform polymer, monodisperse polymer	A polymer composed of molecules uniform with respect to relative molecular mass (see Definition 1.1) and constitution. *Notes* 1. A polymer comprising a mixture of linear and branched chains, all of uniform relative molecular mass, is not uniform. 2. A copolymer comprising linear molecules of uniform relative molecular mass and uniform elemental composition, but different sequence arrangement of the various types of monomeric units, is not uniform (e.g. a copolymer comprising molecules with random arrangement as well as block arrangement of monomeric units). 3. A polymer uniform with respect only to either relative molecular mass or constitution may be termed 'uniform', provided a suitable qualifier is used (e.g. 'a polymer uniform with respect to relative molecular mass'). 4. The adjectives 'monodisperse' and 'polydisperse' (see Definition

Term	*Definition*
	2.4) are deeply rooted in the literature despite being non-descriptive and self-contradictory. They are in common usage and it is recognized that they will continue to be used for a certain time, nevertheless more satisfactory terms are clearly desirable. After an extensive search for possible replacements, the new terms 'uniform' and 'non-uniform' (see Definition 2.4) have been selected and they are now the preferred adjectives.
2.4 non-uniform polymer, polydisperse polymer	A polymer comprising molecules non-uniform with respect to relative molecular mass (see Definition 1.1) or constitution or both. *Note* See Definition 2.3, *Note* 3.
2.5 molar-mass average, relative molecular-mass average/molecular-weight average	recommended symbols: $\bar{M}_{\mathrm{k}}$ and $\bar{M}_{\mathrm{r,k}}$, respectively In both symbols, k specifies the type of average. Any average of the molar mass (see Definition 1.2) or relative molecular mass (molecular weight) (see Definition 1.1) for a non-uniform polymer. *Notes* 1. An infinite number of molar-mass averages can in principle be defined, but only a few types of averages are directly accessible experimentally. The most important averages are defined by simple moments of the distribution functions (see Definition 2.13) and are obtained by methods applied to systems in thermodynamic equilibrium, such as osmometry, light scattering and sedimentation equilibrium. Hydrodynamic methods, as a rule, yield more complex molar-mass averages. 2. Any molar-mass average can be defined in terms of mass fractions or mole fractions. In this document only a few of the important molar-mass averages are given in terms of the mass fractions, w_i, of the species with molar mass M_i. These definitions are most closely related to the experimental determination of molar-mass averages.
2.6 number-average molar mass	recommended symbol: $\bar{M}_{\mathrm{n}}$ $$\bar{M}_{\mathrm{n}} = \frac{1}{\sum_i (w_i/M_i)}$$
number-average relative molecular mass, number-average molecular weight	recommended symbol: $\bar{M}_{\mathrm{r,n}}$ $$\bar{M}_{\mathrm{r,n}} = \frac{1}{\sum_i (w_i/M_{\mathrm{r},i})}$$ For explanation of symbols, see Definition 2.5.

Term	*Definition*
2.7 mass-average molar mass	recommended symbol: $\bar{M}_w$ $\bar{M}_w = \sum_i w_i M_i$
mass-average relative molecular mass, weight-average molecular weight	recommended symbol: $\bar{M}_{r,w}$ $\bar{M}_{r,w} = \sum_i w_i M_{r,i}$ For explanation of symbols, see Definition 2.5.
2.8 z-average molar mass	recommended symbol: $\bar{M}_z$ $\bar{M}_z = \frac{\sum_i w_i M_i^2}{\sum_i w_i M_i}$
z-average relative molecular mass, z-average molecular weight	recommended symbol: $\bar{M}_{r,z}$ $\bar{M}_{r,z} = \frac{\sum_i w_i M_{r,i}^2}{\sum_i w_i M_{r,i}}$ For explanation of symbols, see Definition 2.5.
2.9 (z+1)-average molar mass	recommended symbol: $\bar{M}_{z+1}$ $\bar{M}_{z+1} = \frac{\sum_i w_i M_i^3}{\sum_i w_i M_i^2}$
(z+1)-average relative molecular mass, (z+1)-average molecular weight	recommended symbol: $\bar{M}_{r,z+1}$ $\bar{M}_{r,z+1} = \frac{\sum_i w_i M_{r,i}^3}{\sum_i w_i M_{r,i}^2}$ For explanation of symbols, see Definition 2.5.
2.10 viscosity-average molar mass	recommended symbol: $\bar{M}_v$ $\bar{M}_v = \left[\sum_i w_i M_i^a\right]^{1/a}$
viscosity-average relative molecular mass, viscosity-average molecular weight	recommended symbol: $\bar{M}_{r,v}$ $\bar{M}_{r,v} = \left[\sum_i w_i M_{r,i}^a\right]^{1/a}$ where a is the exponent in the Mark–Houwink equation (see Definition 3.2.25), $[\eta] = K \cdot M^a$; for explanation of the other symbols, see Definition 2.5.

Term	*Definition*
	Note The exponent a is not identical with the adjustable parameter of some of the distribution functions (see Definitions 2.16–2.20) or with the persistence length (see Definition 1.22).
2.11 apparent molar mass	recommended symbol: M_{app}
apparent relative molecular mass, apparent molecular weight	recommended symbol: $M_{\mathrm{r,app}}$ Molar mass, relative molecular mass, or molecular weight calculated from experimental data without the application of appropriate corrections, such as for finite polymer concentration, association, preferential solvation, compositional heterogeneity (see Definition 2.1), constitutional heterogeneity (see Definition 2.2).
2.12 average degree of polymerization	recommended symbol: $\bar{X}_{\mathrm{k}}$ where k specifies the type of average. Any average of the degree of polymerization (see Definition 1.3) for a polymer. *Note* Definitions 2.5–2.10 apply directly to averages of the degree of polymerization when X is substituted for M in the formulae.
2.13 distribution function	A normalized function giving the relative amount of a portion of a polymeric substance with a specific value, or a range of values, of a random variable or variables. *Notes* 1. Distribution functions may be discrete, i.e. take on only certain specified values of the random variable(s), or continuous, i.e. take on any intermediate value of the random variable(s), in a given range. Most distributions in polymer science are intrinsically discrete, but it is often convenient to regard them as continuous or to use distribution functions that are inherently continuous. 2. Distribution functions may be integral (or cumulative), i.e. give the proportion of the population for which a random variable is less than or equal to a given value. Alternatively they may be differential distribution functions (or probability density functions), i.e. give the (maybe infinitesimal) proportion of the population for which the random variable(s) is (are) within a (maybe infinitesimal) interval of its (their) range(s). 3. Normalization requires that: (i) for a discrete differential distribution function, the sum of the function values over all possible values of the random variable(s) be unity; (ii) for a continuous differential distribution function, the integral over the entire range of the random variable(s) be unity; (iii) for an integral (cumulative) distribution function, the function value at the upper limit of the random variable(s) be unity.

Term	*Definition*
2.14 number-distribution function	A distribution function (see Definition 2.13) in which the relative amount of a portion of a substance with a specific value, or a range of values, of the random variable(s) is expressed in terms of mole fraction.
2.15 mass-distribution function, weight-distribution function	A distribution function (see Definition 2.13) in which the relative amount of a portion of a substance with a specific value, or a range of values, of the random variable(s) is expressed in terms of mass fraction.
2.16 Schulz–Zimm distribution	A continuous distribution with the differential mass-distribution function (see Definition 2.15) of the form $f_w(w)\,dx = \frac{a^{b+1}}{\Gamma(b+1)} x^b \exp(-ax)\,dx$, where x is a parameter characterizing the chain length, such as relative molecular mass (see Definition 1.1) or degree of polymerization (see Definition 1.3), a and b are positive adjustable parameters, and $\Gamma(b+1)$ is the gamma function of $(b+1)$.
2.17 most probable distribution	A discrete distribution with the differential mass-distribution function (see Definition 2.15) of the form $f_w(x) = a^2 x(1-a)^{x-1}$. For explanation of symbols, see Definition 2.16. *Notes* 1. For large values of x, the most probable distribution converges to the particular case of the Schulz–Zimm distribution (see Definition 2.16) with $b=1$. 2. In the literature, this distribution is sometimes referred to as the Flory distribution or the Schulz–Flory distribution.
2.18 Poisson distribution	A discrete distribution with the differential mass-distribution function (see Definition 2.15) of the form $f_w(x) = \frac{x}{a+1}\,\frac{e^{-a}a^{x-1}}{(x-1)!}$ For explanation of symbols, see Definition 2.16.
2.19 Tung distribution	A continuous distribution with the differential mass-distribution function (see Definition 2.15) of the form $f_w(x)\,dx = abx^{b-1}\exp(ax^b)\,dx$. For explanation of symbols, see Definition 2.16.

Term	Definition
2.20 logarithmic normal distribution	A continuous distribution with the differential mass-distribution function (see Definition 2.15) of the form $$f_{\mathrm{w}}(x)\,\mathrm{d}x = \frac{1}{a\sqrt{\pi x}}\exp\left(-\frac{1}{a^2}\ln^2\frac{x}{b}\right)\mathrm{d}x.$$ For explanation of symbols, see Definition 2.16.
2.21 polymolecularity correction	A correction applied to relationships between a property and the molar mass (see Definition 1.2) or relative molecular mass (see Definition 1.1), obtained from polymers non-uniform (see Definition 2.4) with respect to relative molecular mass, in order to obtain the corresponding relationship for polymers strictly uniform (see Definition 2.3) with respect to relative molecular mass.

3 DILUTE POLYMER SOLUTIONS

3.1 General and thermodynamic terms

Term	Definition
3.1.1 dilute solution	A solution in which the sum of the volumes of the domains occupied by the solute molecules or particles is substantially less than the total volume of the solution. *Note* The term 'domain' refers to the smallest convex body that contains the molecule or particle in its average shape.
3.1.2 cross-over concentration	recommended symbol: c^* The concentration range at which the sum of the volumes of the domains occupied by the solute molecules or particles in solution is approximately equal to the total volume of that solution. *Note* For the meaning of the term 'domain', see the note in Definition 3.1.1.
3.1.3 polymer–solvent interaction	The sum of the effects of all intermolecular interactions between polymer and solvent molecules in solution that are reflected in the Gibbs and Helmholtz energies of mixing.
3.1.4 quality of solvent, thermodynamic quality of solvent	Qualitative characterization of the polymer–solvent interaction (see Definition 3.1.3). A solution of a polymer in a 'better' solvent is characterized by a higher value of the second virial coefficient (see Definition 3.1.7) than a solution of the same polymer in a 'poorer' solvent.
3.1.5 theta state	The state of a polymer solution for which the second virial coefficient (see Definition 3.1.7) is zero.

Term	Definition
	Notes 1. In some respects, a polymer solution in the theta state resembles an ideal solution and the theta state may be referred to as a pseudo-ideal state. However, a solution in the theta state must not be identified with an ideal solution. 2. In the literature, the theta state is also written 'θ state'. 3. The solvent involved is often referred to as 'theta solvent'. 4. It is assumed that the molar mass of the polymer is high.
3.1.6 theta temperature	The temperature at which a solution is in the theta state (see Definition 3.1.5). *Note* In the literature, the theta temperature is also written 'θ temperature'.
3.1.7 virial coefficients, virial coefficients of the chemical potential	recommended symbol: A_i, where $i=1$, 2, etc. The coefficients in the expansion of the chemical potential of the solvent, μ_s, in powers of the mass concentration, c, of the solute, i.e. $\mu_s - \mu_s^0 = -\pi V_s = -RTV_s(A_1c + A_2c^2 + A_3c^3 + \cdots)$, where μ_s^0 is the chemical potential of the solvent in the reference state at the temperature of the system and ambient pressure, π is the osmotic pressure and V_s is the partial molar volume of the solvent. In solvents comprising more than one component, the definition applies to any solvent component. The first virial coefficient is the reciprocal number-average molar mass (see Definition 2.6), i.e. $A_1 = 1/\bar{M}_n$. The second and higher virial coefficients, A_2, A_3, ..., respectively, describe polymer–solvent (see Definition 3.1.3) and polymer–polymer interactions. *Note* The factor RT is sometimes included in the virial coefficients.
3.1.8 excluded volume of a segment	The volume from which a segment of a macromolecule in solution effectively excludes all other segments, i.e. those belonging to the same macromolecule as well as those belonging to other macromolecules. *Note* The excluded volume of a segment depends on the Gibbs and Helmholtz energies of mixing of solvent and polymer, i.e. on the thermodynamic quality of the solvent (see Definition 3.1.4), and is not a measure of the geometrical volume of that segment.
3.1.9 excluded volume of a macromolecule	The volume from which a macromolecule in a dilute solution effectively excludes all other macromolecules.

	Term	Definition
		Note The excluded volume of a macromolecule depends on the Gibbs and Helmholtz energies of mixing of solvent and polymer, i.e. on the thermodynamic quality of the solvent (see Definition 3.1.4), and is not a measure of the geometrical volume of that macromolecule.
3.1.10	expansion factor	recommended symbols: α_r, α_s, α_η The ratio of a dimensional characteristic of a macromolecule in a given solvent at a given temperature to the same dimensional characteristic in the theta state (see Definition 3.1.5) at the same temperature. The most frequently used expansion factors are: expansion factor of the mean-square end-to-end distance (see Definition 1.12), $\alpha_r = (\langle r^2 \rangle / \langle r^2 \rangle_0)^{1/2}$; expansion factor of the radius of gyration (see Definition 1.9), $\alpha_s = (\langle s^2 \rangle / \langle s^2 \rangle_0)^{1/2}$; viscosity expansion factor $\alpha_\eta = ([\eta]/[\eta]_\theta)^{1/3}$, where $[\eta]$ and $[\eta]_\theta$ are the intrinsic viscosity (see Definition 3.2.21) in a given solvent and in the theta state at the same temperature, respectively. *Note* Expansion factors defined by different dimensional characteristics are not exactly equal, nor need they have a constant ratio as a function of relative molecular mass.
3.1.11	Flory–Huggins theory	A thermodynamic theory of polymer solutions, first formulated independently by Flory and by Huggins, in which the thermodynamic quantities of the solution are derived from a simple concept of combinatorial entropy of mixing and a reduced Gibbs-energy parameter, the 'χ parameter' (see Definition 3.1.12).
3.1.12	χ parameter	recommended symbol: χ A numerical parameter employed in the Flory–Huggins theory (see Definition 3.1.11), which accounts in the main for the contribution of the non-combinatorial entropy of mixing and for the enthalpy of mixing.
3.1.13	preferential sorption, selective sorption	An equilibrium phenomenon, operative in polymer solutions in multicomponent solvents and in polymer networks (see Definition 1.26) swollen by multicomponent solvents, that produces differences in solvent composition in the polymer-containing region and in the pure solvent which is in thermodynamic equilibrium with that region.
3.1.14	selective solvent	A medium that is a solvent for at least one component of a mixture of polymers, or for at least one block of a block or graft polymer, but a non-solvent for the other component(s) or block(s).

Term	*Definition*
3.1.15 co-solvency	The dissolution of a polymer in a solvent comprising more than one component, each component of which by itself is a non-solvent for the polymer.
3.1.16 solubility parameter (of a polymer)	recommended symbol: δ A characteristic of a polymer used in predicting solubility of that polymer in a given solvent. *Notes* 1. For a substance of low molecular weight, the value of the solubility parameter is often estimated from the enthalpy of vaporization; for a polymer, it is usually taken to be the value of the solubility parameter of the solvent producing the solution with maximum intrinsic viscosity (see Definition 3.2.21) or maximum swelling of a network (see Definition 1.26) of the polymer. 2. The solubility parameter is usually expressed in $(\mathrm{cal/cm^3})^{1/2}$ or, preferably, $(\mathrm{J/cm^3})^{1/2}$ units. The units must always be given; 1 $(\mathrm{cal/cm^3})^{1/2} \approx 2.05$ $(\mathrm{J/cm^3})^{1/2}$.
3.1.17 isopycnic	An adjective describing components of a multicomponent system with equal partial specific volumes.

3.2 Transport properties

Term	*Definition*
3.2.1 frictional coefficient	recommended symbol: f A tensor correlating the frictional force F, opposing the motion of a particle in a viscous fluid and the velocity u of this particle relative to the fluid. *Note* In the case of an isolated spherical particle in a viscous isotropic fluid, f is a constant and $F = fu$.
3.2.2 hydrodynamically equivalent sphere	A hypothetical sphere, impenetrable to the surrounding medium, displaying in a hydrodynamic field the same frictional effect as an actual polymer molecule. *Note* The size of a hydrodynamically equivalent sphere may be different for different types of motion of the macromolecule, e.g. for diffusion and for viscous flow.
3.2.3 hydrodynamic volume	The volume of a hydrodynamically equivalent sphere (see Definition 3.2.2).
3.2.4 bead-rod model	A model simulating the hydrodynamic properties of a chain

Term	*Definition*
	macromolecule consisting of a sequence of beads, each of which offers hydrodynamic resistance to the flow of the surrounding medium and is connected to the next bead by a rigid rod which does not. The mutual orientation of the rods is random.
3.2.5 bead-spring model	A model simulating the hydrodynamic properties of a chain macromolecule consisting of a sequence of beads, each of which offers hydrodynamic resistance to the flow of the surrounding medium and is connected to the next bead by a spring which does not contribute to the frictional interaction but which is responsible for the elastic and deformational properties of the chain. The mutual orientation of the springs is random.
3.2.6 freely draining	An adjective referring to a chain macromolecule the segments of which produce such small frictional effects when moving in a medium that the hydrodynamic field in the vicinity of a given segment is not affected by the presence of other segments. Thus, the solvent can flow virtually undisturbed through the domain occupied by a freely draining macromolecule.
3.2.7 non-draining	An adjective describing a chain macromolecule that behaves in a hydrodynamic field as though the solvent within the domain of the macromolecule were virtually immobilized with respect to the macromolecule.
3.2.8 partially draining	An adjective describing a chain macromolecule that behaves in a hydrodynamic field as though the solvent within the domain of the macromolecule were progressively more immobilized with respect to the macromolecule in the direction from its outer fringes inward. *Note* A freely draining (see Definition 3.2.6) macromolecule and a non-draining (see Definition 3.2.7) macromolecule are two extremes of the concept of a partially draining macromolecule.
3.2.9 streaming birefringence, flow birefringence	The birefringence induced by flow in liquids, solutions and dispersions of optically anisotropic, anisometric or deformable molecules or particles due to a non-random orientation of the molecules or particles.
3.2.10 rotational diffusion	A process by which the equilibrium statistical distribution of the overall orientation of molecules or particles is maintained or restored. *Note* Rotational diffusion may be compared to translational diffusion,

Term		*Definition*
		through which the equilibrium statistical distribution of position in space is maintained or restored.
3.2.11	sedimentation coefficient	recommended symbol: s A parameter characterizing the motion of a particle in a centrifugal field and defined as the velocity of motion u due to unit centrifugal acceleration, i.e. $s = u/(r \cdot \omega^2)$, where ω is the angular velocity and r the distance from the centre of rotation. *Note* The unit 10^{-13} second is useful; this unit has been referred to as a 'svedberg' which, however, is not an SI unit.
3.2.12	sedimentation equilibrium	The equilibrium established in a centrifugal field when there is no net flux of any component across any plane perpendicular to the centrifugal force.
3.2.13	equilibrium sedimentation (method)	A method by which the distribution of the concentration of the solute or dispersed component in a dilute solution or dispersion along the centrifuge cell is measured at sedimentation equilibrium (see Definition 3.2.12), and the results are interpreted in terms of molar masses (see Definition 1.2) or their distribution, or both.
3.2.14	sedimentation velocity method	A method by which the velocity of motion of solute component(s) or dispersed particles is measured and the result is expressed in terms of its (their) sedimentation coefficient(s) (see Definition 3.2.11).
3.2.15	Archibald's method	A sedimentation method based on the fact that at the meniscus and at the bottom of the centrifuge cell there is never a flux of the solute across a plane perpendicular to the radial direction and the equations characterizing the sedimentation equilibrium (see Definition 3.2.12) always apply there, even though the system as a whole may be far from equilibrium. *Note* The use of the term 'approach to sedimentation equilibrium' for Archibald's method is discouraged, since it has a more general meaning.
3.2.16	equilibrium sedimentation in a density gradient	An equilibrium sedimentation technique (see Definition 3.2.13) working with a multi-component solvent forming a density gradient in a centrifugal field.
3.2.17	relative viscosity, viscosity ratio	recommended symbol: η_r The ratio of the viscosity of the solution, η, to the viscosity of the solvent, η_s, i.e. $\eta_r = \eta/\eta_s$.

Term — *Definition*

3.2.18 relative viscosity increment

recommended symbol: η_i

The ratio of the difference between the viscosities of solution and solvent to the viscosity of the solvent, i.e. $\eta_i = (\eta - \eta_s)/\eta_s$. For explanation of symbols, see Definition 3.2.17.

Note

The use of the term 'specific viscosity' for this quantity is discouraged, since the relative viscosity increment does not have the attributes of a specific quantity.

3.2.19 reduced viscosity, viscosity number

The ratio of the relative viscosity increment (see Definition 3.2.18) to the mass concentration of the polymer, c, i.e. η_i/c.

Notes

1. The unit must be specified; cm^3/g is recommended.
2. This quantity and those in Definitions 3.2.20 and 3.2.21 are neither viscosities nor pure numbers. The terms are to be looked on as traditional names. Any replacement by consistent terminology would produce unnecessary confusion in the polymer literature.

3.2.20 inherent viscosity, logarithmic viscosity number

recommended symbols: η_{inh}, η_{ln}, respectively

The ratio of the natural logarithm of the relative viscosity (see Definition 3.2.17) to the mass concentration of the polymer, c, i.e.

$$\eta_{inh} \equiv \eta_{ln} = (\ln \eta_r)/c.$$

Notes

See notes under Definition 3.2.19.

3.2.21 intrinsic viscosity, limiting viscosity number

recommended symbol: $[\eta]$

The limiting value of the reduced viscosity (see Definition 3.2.19) or the inherent viscosity (see Definition 3.2.20) at infinite dilution of the polymer, i.e.

$$[\eta] = \lim_{c \to 0} (\eta_i/c) = \lim_{c \to 0} \eta_{inh}$$

Notes

1. This term is also known in the literature as the Staudinger index.
2. See notes under Definition 3.2.19.

3.2.22 Huggins equation

The equation describing the dependence of the reduced viscosity (see Definition 3.2.19), η_i/c, on the mass concentration of the polymer, c, for dilute polymer solutions (see Definition 3.1.1) of the form

$$\eta_i/c = [\eta] + k_H[\eta]^2 c$$

Term		Definition
		where k_H is the Huggins coefficient (see Definition 3.2.23) and $[\eta]$ is the intrinsic viscosity (see Definition 3.2.21).
3.2.23	Huggins coefficient	recommended symbol: k_H A parameter in the Huggins equation (see Definition 3.2.22).
3.2.24	viscosity function	recommended symbol: Φ A coefficient connecting the intrinsic viscosity (see Definition 3.2.21), the radius of gyration (see Definition 1.9) and the molar mass (see Definition 1.2) of a chain macromolecule, according to the equation $[\eta]=\Phi\, 6^{3/2}\langle s^2\rangle^{3/2}/M$. *Note* The viscosity function is often referred to as the Flory constant.
3.2.25	Mark–Houwink equation	The equation describing the dependence of the intrinsic viscosity (see Definition 3.2.21) of a polymer on its relative molecular mass (molecular weight) (see Definition 1.1) and having the form $[\eta]=K\cdot M_r^a$ where K and a are constants, the values of which depend on the nature of the polymer and solvent as well as on temperature; M_r is usually one of the relative molecular-mass averages (see Definition 2.5). *Notes* 1. The use of this equation with the relative molecular mass (molecular weight) is recommended, rather than with molar mass (which has the dimension of mass divided by amount of substance), since in the latter case the constant K assumes awkward and variable dimensions owing to the fractional and variable nature of the exponent a. 2. Kuhn and Sakurada have also made important contributions and their names are sometimes included, as, for example, in the Kuhn–Mark–Houwink–Sakurada equation.

3.3 Coherent elastic scattering of radiation

A beam of radiation traversing a medium may be attenuated and partially scattered. The definitions below are for those cases in which the attenuation of the incident beam is due only to scattering, the energy of scattering quanta is the same as that of quanta in the primary beam (elastic scattering) and phase relationships between independent scatterers are retained (coherent scattering). This document deals with light scattering (LS), small-angle x-ray scattering (SAXS), and small-angle neutron scattering (SANS). In light scattering the polarization of light is relevant; plane-polarized light is considered here only, and it is called vertically polarized (v) if the electric vector of the beam is perpendicular to the plane containing the source, sample and detector, and horizontally polarized (h) if the electric vector lies in that plane. Unpolarized light is considered to be a mixture of equal parts of v and h light.

Term		*Definition*
3.3.1	small particle	A particle much smaller than the wavelength of the radiation in the medium. In practice, all dimensions of a particle considered small must be less than about one-twentieth of the wavelength employed.
3.3.2	large particle	A particle with dimensions comparable with the wavelength of the radiation in the medium or larger. In practice a particle must be treated as large, if its largest dimension exceeds about one-twentieth of the wavelength employed.
3.3.3	scattering angle, angle of observation	recommended symbol: θ The angle between the forward direction of the incident beam and a straight line connecting the scattering point and the detector.
3.3.4	scattering vector	Vector difference between the wave propagation vectors of the incident and the scattered beam, both of length $2\pi/\lambda$, where λ is the wavelength of the scattered radiation in the medium.
3.3.5	length of the scattering vector	recommended symbol: q The length of the scattering vector (see Definition 3.3.4) is $q = (4\pi/\lambda) \cdot \sin(\theta/2)$, where λ is the wavelength of the scattered radiation in the medium and θ is the scattering angle (see Definition 3.3.3).
3.3.6	refractive index increment	recommended symbol: $\partial n/\partial C$ The change of the solution refractive index, n, with solute concentration, C. *Notes* 1. The solute concentration is most frequently expressed in terms of mass concentration, molality or volume fraction. If expressed in terms of mass concentration or molality, the corresponding refractive index increments are referred to as specific or molal refractive index increments, respectively. 2. Following use of the full name, the abbreviated name refractive increment may be used.
3.3.7	Rayleigh ratio	recommended symbols: $R(\theta)$, R_θ The quantity used to characterize the scattered intensity at the scattering angle θ (see Definition 3.3.3), defined as $R(\theta) = i_\theta r^2/(I \cdot f \cdot V)$, where I is the intensity of the incident radiation, i_θ is the total intensity of scattered radiation observed at an angle θ and a distance r from the point of scattering and V is the scattering volume. The factor f takes account of polarization phenomena.

Term	*Definition*
	It depends on the type of radiation employed: 1. for light scattering, dependent on the polarization of the incident beam, $f=1$ for vertically polarized light, $f=\cos^2\theta$ for horizontally polarized light, $f=(1+\cos^2\theta)/2$ for unpolarized light; 2. for small-angle neutron scattering $f=1$; 3. for small-angle x-ray scattering $f\approx 1$, if $\theta<$ ca. $5°$. *Notes* 1. The dimension of $R(\theta)$ is $(\text{length})^{-1}$ and the unit commonly used is cm^{-1}. 2. In small-angle neutron scattering the term cross-section is often used instead of $R(\theta)$; the two quantities are identical.
3.3.8 excess Rayleigh ratio	The difference between the Rayleigh ratio (see Definition 3.3.7) for a dilute solution and for pure solvent. *Note* If the scattering intensity is not reduced to the Rayleigh ratio, the difference between the scattering intensities for a dilute solution and that for pure solvent is named 'excess scattering'.
3.3.9 turbidity	recommended symbol: τ The apparent absorbance of the incident radiation due to scattering. *Note* For small particles (see Definition 3.3.1), direct proportionality exists between turbidity and the Rayleigh ratio (see Definition 3.3.7).
3.3.10 particle scattering function, particle scattering factor	recommended symbols: $P(\theta)$, P_θ The ratio of the intensity of radiation scattered at an angle of observation θ (see Definition 3.3.3) to the intensity of radiation scattered at an angle zero, i.e. $P(\theta)\equiv R(\theta)/R(0)$.
3.3.11 Zimm plot	A diagrammatic representation of data on scattering from large particles (see Definition 3.3.2), corresponding to the equation $$\frac{Kc}{\Delta R(\theta)}=\frac{1}{\bar{M}_{\mathrm{w}}P(\theta)}+2A_2c+\cdots,$$ and used for the simultaneous evaluation of the mass-average molar mass (see Definition 2.7) $\bar{M}_{\mathrm{w}}$, the second virial coefficient of the chemical potential (see Definition 3.1.7) A_2 and (usually) the z-average radius of gyration (see Definition 1.9) $\langle s^2\rangle_z^{1/2}$; c is the mass concentration of the solute, $\Delta R(\theta)$ the excess Rayleigh ratio (see Definition 3.3.8), and $P(\theta)$ the particle

Term	Definition
	scattering function (see Definition 3.3.10) that comprises (usually) the z-average radius of gyration. K depends on the solute, the temperature and the type of radiation employed. *Note* Several modifications of the Zimm plot are in frequent use; the most common one uses the excess scattering (see Definition 3.3.8) instead of the excess Rayleigh ratio.
3.3.12 Guinier plot	A diagrammatic representation of data on scattering from large particles (see Definition 3.3.2), obtained at different angles but at the same concentration, constructed by plotting $\log \Delta R(\theta)$ or $\log P(\theta)$ versus $\sin^2(\theta/2)$ or q^2, and (usually) used for the evaluation of the radius of gyration (see Definition 1.9). $\Delta R(\theta)$ is the excess Rayleigh ratio (see Definition 3.3.8), $P(\theta)$ the particle scattering function (see Definition 3.3.10), θ the scattering angle (see Definition 3.3.3) and q the length of the scattering vector (see Definition 3.3.5).
3.3.13 Kratky plot	A diagrammatic representation of scattering data on large particles, obtained at different angles but at the same concentration, constructed by plotting $\sin^2(\theta/2)\cdot\Delta R(\theta)$ versus $\sin(\theta/2)$, or $q^2\Delta R(\theta)$ versus q, and used for the determination of molecular shape. For explanation of symbols, see Definition 3.3.12.
3.3.14 dissymmetry of scattering	recommended symbol: $z(\theta_1, \theta_2)$ The ratio of two Rayleigh ratios (see Definition 3.3.7) for different angles of observation (see Definition 3.3.3), i.e. $z(\theta_1, \theta_2) = R(\theta_1)/R(\theta_2), \theta_1 < \theta_2$. *Note* The angles must be specified; in light scattering it is customary to let $\theta_2 = 180° - \theta_1$ and, most frequently, $\theta_1 = 45°$ and $\theta_2 = 135°$.
3.3.15 depolarization of scattered light	The phenomenon, due primarily to the anisotropy of the polarizability of the scattering medium, resulting from the fact that the electric vectors of the incident and scattered beams are not coplanar and that, therefore, light scattered from a vertically (horizontally) polarized incident beam contains a horizontal (vertical) component.
3.3.16 turbidimetric titration	The process in which a precipitant is added incrementally to a highly dilute polymer solution and the intensity of light scattered by, or the turbidity (see Definition 3.3.9) due to, the finely dispersed particles of the polymer-rich phase (see Definition 3.4.3) is measured as a function of the amount of precipitant added.

Term	Definition
3.3.17 isorefractive	An adjective describing components of a multicomponent system having zero refractive index increments with respect to each other.
3.3.18 Mie scattering	Scattering of light by particles with size larger than approximately one-half of the wavelength of incident light. *Note* For homogeneous spheres, this phenomenon is rigorously described by the theory developed by Mie.

3.4 Separation

Term	Definition
3.4.1 fractionation	A process by means of which macromolecular species differing in some characteristic (chemical composition, relative molecular mass (see Definition 1.1), branching, stereoregularity, etc) are separated from each other.
3.4.2 polymer-poor phase, dilute phase	That phase of a two-phase equilibrium system, consisting of a polymer and low-molecular-weight material, in which the polymer concentration is lower. *Note* The use of the name 'sol phase' is discouraged.
3.4.3 polymer-rich phase, concentrated phase	That phase of a two-phase equilibrium system, consisting of a polymer and low-molecular-weight material, in which the polymer concentration is higher. *Note* The use of the name 'gel phase' is discouraged.
3.4.4 precipitation fractionation	A process in which a polymeric material, consisting of macromolecules differing in some characteristic affecting their solubility, is separated from solution into fractions by successively decreasing the solution power of the solvent, resulting in the repeated formation of a two-phase system in which the less soluble components concentrate in the polymer-rich phase (see Definition 3.4.3).
3.4.5 extraction fractionation	A process in which a polymeric material, consisting of macromolecules differing in some characteristic affecting their solubility, is separated from a polymer-rich (see Definition 3.4.3) phase into fractions by successively increasing the solution power of the solvent, resulting in the repeated formation of a two-phase system in which the more soluble components concentrate in the polymer-poor phase (see Definition 3.4.2).

	Term	*Definition*
3.4.6	size-exclusion chromatography	recommended abbreviation: SEC A separation technique in which separation mainly according to the hydrodynamic volume (see Definition 3.2.3) of the molecules or particles takes place in a porous non-adsorbing material with pores of approximately the same size as the effective dimensions in solution of the molecules to be separated.
	gel-permeation chromatography	recommended abbreviation: GPC The term 'gel-permeation chromatography' should only be used when the porous non-adsorbing material is a gel. The term 'size-exclusion chromatography' is preferred.
3.4.7	molar-mass exclusion limit, molecular-weight exclusion limit	The maximum value of the molar mass (see Definition 1.2) or molecular weight (see Definition 1.1) of molecules or particles, in a specific polymer–solvent system, that can enter into the pores of the porous non-adsorbing material used in size-exclusion chromatography (see Definition 3.4.6). *Note* For particles with molar mass or molecular weight larger than the exclusion limit the separation effect of the size-exclusion chromatography vanishes.
3.4.8	elution volume	The volume of a solvent passed, since the injection of the sample, through a size-exclusion chromatography (see Definition 3.4.6) bed at the time at which a specified signal of the detector has been recorded.
3.4.9	retention volume	The elution volume (see Definition 3.4.8) at the maximum concentration of an elution peak.
3.4.10	universal calibration	A calibration of a size-exclusion chromatography (see Definition 3.4.6) set-up based on the finding that the retention volume (see Definition 3.4.9) of a molecular or particulate species is a single-valued function of an appropriate size parameter of this molecule or particle, irrespective of its chemical nature and structure. *Note* The product of the intrinsic viscosity (see Definition 3.2.21) and molar mass (see Definition 1.2), $[\eta]M$, has been widely used as the size parameter.
3.4.11	spreading function	The normalized signal produced, as a function of elution volume (see Definition 3.4.8), at the outlet of a size-exclusion chromatography (see Definition 3.4.6) set-up, by an instantaneous injection of a uniform sample.

Term	*Definition*
3.4.12 plate number	recommended symbol: N A characteristic of the efficiency of a size-exclusion chromatography (see Definition 3.4.6) set-up in terms of band broadening, defined as $N=(V_R/\sigma_V)^2$, where V_R is the retention volume (see Definition 3.4.9) of an individual low-molecular-weight compound, and σ_V is the corresponding half width at 60.7% peak height of the elution peak.
3.4.13 plate height	recommended symbol: H The length of a part of a size-exclusion chromatography (see Definition 3.4.6) bed corresponding to one plate, i.e. the length of the bed, L, divided by its plate number (see Definition 3.4.12), N, or $H=L/N$.

4 ALPHABETICAL INDEX OF TERMS

Term	*Symbol*	*Definition number*
most probable distribution		2.17
network		1.26
non-draining		3.2.7
non-uniform polymer		2.4
number-average degree of polymerization	$\bar{X}_n$	2.12
number-average molar mass	$\bar{M}_n$	2.6
number-average molecular weight	$\bar{M}_{r,n}$	2.6
number-average relative molecular mass	$\bar{M}_{r,n}$	2.6
number-distribution function		2.14
partially draining		3.2.8
particle scattering factor	$P(\theta)$, P_θ	3.3.10
particle scattering function	$P(\theta)$, P_θ	3.3.10
persistence length	a	1.22
perturbed dimensions		1.8
plate height	H	3.4.13
plate number	N	3.4.12
Poisson distribution		2.18
polydisperse polymer		2.4
polymer-poor phase		3.4.2
polymer-rich phase		3.4.3
polymer–solvent interaction		3.1.3
polymolecularity correction		2.21
Porod–Kratky chain		1.21
precipitation fractionation		3.4.4
preferential sorption		3.1.13
probability density function		2.13
quality of solvent		3.1.4
radius of gyration	s, $\langle s^2 \rangle^{1/2}$	1.9
random coil		1.15
random-walk chain		1.16
Rayleigh ratio	$R(\theta)$, R_θ	3.3.7
reduced viscosity		3.2.19
refractive increment	$\partial n/\partial C$	3.3.6
refractive index increment	$\partial n/\partial C$	3.3.6
relative molecular mass	M_r	1.1
relative molecular-mass average	$\bar{M}_r$	2.5
relative viscosity	η_r	3.2.17
relative viscosity increment	η_i	3.2.18
retention volume		3.4.9
root-mean-square end-to-end distance	$\langle r^2 \rangle^{1/2}$	1.12
rotational diffusion		3.2.10
scattering angle	θ	3.3.3
scattering vector		3.3.4
Schulz–Flory distribution		2.17
Schulz–Zimm distribution		2.16
second virial coefficient	A_2	3.1.7
sedimentation coefficient	s	3.2.11
sedimentation equilibrium		3.2.12
sedimentation velocity method		3.2.14
selective sorption		3.1.13
selective solvent		3.1.14
short-chain branch		1.23
short-range interaction		1.5
short-range intramolecular interaction		1.5

Term	*Symbol*	*Definition number*
size-exclusion chromatography		3.4.6
small particle		3.3.1
solubility parameter	δ	3.1.16
specific refractive index increment		3.3.6
spreading function		3.4.11
statistical coil		1.15
statistical segment		1.18
Staudinger index	$[\eta]$	3.2.21
steric factor	σ	1.20
streaming birefringence		3.2.9
thermodynamic quality of solvent		3.1.4
thermodynamically equivalent sphere		1.4
theta solvent		3.1.5
theta state		3.1.5
theta temperature		3.1.6
Tung distribution		2.19
turbidimetric titration		3.3.16
turbidity	τ	3.3.9
uniform polymer		2.3
universal calibration		3.4.10
unperturbed dimensions		1.7
virial coefficients		3.1.7
virial coefficients of the chemical potential	$A_1, A_2, \ldots$	3.1.7
viscosity function	Φ	3.2.24
viscosity number		3.2.19
viscosity ratio	η_r	3.2.17
viscosity-average degree of polymerization	$\bar{X}_v$	2.12
viscosity-average molar mass	$\bar{M}_v$	2.10
viscosity-average molecular weight	$\bar{M}_{r,v}$	2.10
viscosity-average relative molecular mass	$\bar{M}_{r,v}$	2.10
weight-average degree of polymerization	$\bar{X}_w$	2.12
weight-average molecular weight	$\bar{M}_{r,w}$	2.7
weight-distribution function		2.15
worm-like chain		1.21
Zimm plot		3.3.11
z-average degree of polymerization	$\bar{X}_z$	2.12
z-average molar mass	$\bar{M}_z$	2.8
z-average molecular weight	$\bar{M}_{r,z}$	2.8
z-average relative molecular mass	$\bar{M}_{r,z}$	2.8
(z + 1)-average degree of polymerization	$\bar{X}_{z+1}$	2.12
(z + 1)-average molar mass	$\bar{M}_{z+1}$	2.9
(z + 1)-average molecular weight	$\bar{M}_{r,z+1}$	2.9
(z + 1)-average relative molecular mass	$\bar{M}_{r,z+1}$	2.9

REFERENCES

1 IUPAC. Basic definitions of terms relating to polymers 1974. *Pure Appl. Chem.* **40**, 477–491 (1974). Reprinted as Chapter 1, this volume.

2 IUPAC. Stereochemical definitions and notations relating to polymers (recommendations 1980). *Pure Appl. Chem.* **53**, 733–752 (1981). Reprinted as Chapter 2, this volume.

3 IUPAC. Note on the terminology for molar masses in polymer science. *Makromol. Chem.* **185**, Appendix to No. 1 (1984); *J. Polym. Sci., Polym. Lett. Ed.* **22**, 57 (1984); and in other journals.

4: Definitions of Terms Relating to Crystalline Polymers (1988)

PREAMBLE

The recommendations embodied in this document are concerned with the terminology relating to the structure of crystalline polymers and the process of macromolecular crystallization. The document is limited to systems exhibiting crystallinity in the classical sense of three-dimensionally periodic regularity. The recommendations deal primarily with crystal structures that are comprised of essentially rectilinear, parallel-packed polymer chains, and secondarily, with those comprised of so-called globular macromolecules. Since the latter are biological in nature, they are not covered in detail here. In general, macromolecular systems with mesophases are also omitted, but crystalline polymers with conformational disorder are included.

After a listing of some general definitions relating to crystalline polymers (Section 1), the subject is divided into sections dealing, successively, with local structural arrangements at the scale of a few bond lengths (Section 2), morphological aspects (Section 3), molecular conformation within polymer crystals (Section 4) and, finally, kinetic aspects of crystallization (Section 5). An alphabetical index of terms is provided for the convenience of the reader.

This document relies on the basic definitions of terms in polymer science [1]. It is the second in a current series published by the Commission on Macromolecular Nomenclature dealing with definitions of physical and physicochemical terms in the polymer field (for the first, see Reference [2]).

1 GENERAL DEFINITIONS

Term	*Definition*
1.1 crystallinity	The presence of three-dimensional order on the level of atomic dimensions. *Note* Crystallinity may be detected by diffraction techniques, heat-of-fusion measurements, etc. Some amount of disorder within the crystalline region is not incompatible with this concept.
1.2 crystalline polymer	A polymer showing crystallinity.

Prepared by a working group consisting of G. Allegra (Italy), P. Corradini (Italy), H.-G. Elias (USA), P.H. Geil (USA), H.D. Keith (USA) and B. Wunderlich (USA).
Reprinted from *Pure Appl. Chem.* **61**, 769–785 (1989).

Term	Definition
	Notes 1. One- or two-dimensional order leads to mesophase structure. 2. The range of order may be as small as about 2 nm in one (or more) crystallographic direction(s) and is usually below 50 nm in at least one direction.
1.3 degree of crystallinity	recommended symbols: w_c for mass fraction; ϕ_c for volume fraction. The fractional amount of crystallinity in the polymer sample.

Notes

1. The assumption is made that the sample can be subdivided into a crystalline phase and an amorphous phase (the so-called two-phase model).
2. Both phases are assumed to have properties identical with those of their ideal states, with no influence of interfaces.
3. The degree of crystallinity may be expressed either as the mass fraction or as the volume fraction, the two quantities being related by

$$w_c = \phi_c \rho_c / \rho \qquad (1)$$

where ρ and ρ_c are the densities of the entire sample and of the crystalline fraction, respectively.

4. The degree of crystallinity can be determined by several experimental techniques; among the most commonly used are: (i) x-ray diffraction, (ii) calorimetry, (iii) density measurements, and (iv) infra-red spectroscopy (IR). Imperfections in crystals are not easily distinguished from the amorphous phase. Also, the various techniques may be affected to different extents by imperfections and interfacial effects. Hence, some disagreement among the results of quantitative measurements of crystallinity by different methods is frequently encountered.
5. The following expressions for $w_{c,\alpha}$ are recommended, where the subscript α specifies the particular experimental method used.

(*i*) *By x-ray diffraction:* the degree of crystallinity, $w_{c,x}$, is given by

$$w_{c,x} = I_c/(I_c + K_x I_a) \qquad (2)$$

where I_c and I_a are the integrated intensities scattered over a suitable angular interval by the crystalline and the amorphous phases, respectively, and K_x is a calibration constant. If the sample is anisotropic, a suitable average of the diffracted intensity in reciprocal space must be obtained.

(*ii*) *By calorimetry:* the degree of crystallinity, $w_{c,h}$, is given by

$$w_{c,h} = \Delta h_{fus}/\Delta h_{fus,c} \qquad (3)$$

where Δh_{fus} is the specific enthalpy of fusion of the sample and $\Delta h_{fus,c}$ is the specific enthalpy of fusion of the completely crystalline polymer over the same temperature range. The value of $\Delta h_{fus,c}$ may be obtained by extrapolating Δh_{fus} to the density

Term	Definition

of the completely crystalline polymer, which in turn may be obtained from x-ray diffraction data. The specific enthalpies of fusion are temperature-dependent.

(*iii*) *By density measurements:* the degree of crystallinity, $w_{c,d}$, is given by

$$w_{c,d} = \frac{\rho_c}{\rho} \cdot \frac{\rho - \rho_a}{\rho_c - \rho_a} \qquad (4)$$

where ρ, ρ_c and ρ_a are the densities of the sample, of the completely crystalline polymer and of the completely amorphous polymer, respectively.

(*iv*) *By infra-red spectroscopy:* the degree of crystallinity, $w_{c,i}$, is given by

$$w_{c,i} = \frac{1}{a_c \rho l} \log_{10}(I_0/I) \qquad (5)$$

where I_0 and I are, respectively, the incident and the transmitted intensities at the frequency of the absorption band due to the crystalline portion, a_c is the absorptivity of the crystalline material and l is the thickness of the sample.

1.4 (polymer) crystal — A crystalline domain usually limited by well-defined boundaries.

Notes

1. Polymer crystals frequently do not display the perfection that is usual for low-molar-mass substances.
2. Twinned polymer crystals are, sometimes, erroneously referred to as 'crystals'.
3. Polymer crystals that can be manipulated individually are often called (polymer) single crystals. A single crystal may contain different fold domains.

1.5 (polymer) crystallite — A small crystalline domain.

Notes

1. A crystallite may have irregular boundaries and parts of its constituent macromolecules may extend beyond its boundaries.
2. This definition is not identical with that used in classical crystallography.

1.6 unit cell — The smallest, regularly repeating material portion contained in a parallelepiped from which a crystal is formed by parallel displacements in three dimensions [3].

Notes

1. Unlike in the case of low-molar-mass substances, the unit cell of polymer crystals usually comprises only parts of the polymer

Term	Definition
	molecules and the regularity of the periodic repetition may be imperfect. 2. In the case of parallel-chain crystals, the chain axis is usually denoted by *c* or, sometimes, *b*. 3. This definition applies to the so-called primitive unit cell. In practice, the effective unit cell may consist of more than one primitive unit cell.
1.7 molecular conformation	The conformation of the macromolecule as a whole. *Notes* 1. In the polymer literature, molecular conformation is sometimes referred to as macroconformation. 2. In molecular conformations involving parallel stems, the latter may be confined to the same crystal or may also extend over several crystals.
1.8 local conformation	The conformation of a macromolecule at the scale of the constitutional units. *Note* In the polymer literature, local conformation is sometimes referred to as microconformation.

2 TERMINOLOGY RELATING TO LOCAL CONFORMATION AND STRUCTURAL ASPECTS [4]

Term	Definition
2.1 chain axis	The straight line parallel to the direction of chain extension, connecting the centres of mass of successive blocks of chain units, each of which is contained within an identity period (see Fig. 4.1).
2.2 (chain) identity period, or (chain) repeating distance	The shortest distance along the chain axis for translational repetition of the chain structure. *Notes* 1. The chain identity period is usually denoted by *c*. 2. An example is given by Fig. 4.1.
2.3 (chain) conformational repeating unit	The smallest structural unit of a polymer chain with a given conformation that is repeated along that chain through symmetry operations [5].
2.4 geometrical equivalence	The symmetry correspondence among units belonging to the same chain.

Terms	*Definition*
	Note The symmetry elements always bear a special relationship to the chain axis (see also the note, Definition 2.9).
2.5 equivalence postulate	The working hypothesis that the chain monomeric units are geometrically equivalent [6].
2.6 helix	The molecular conformation of a spiral nature, generated by regularly repeating rotations around the backbone bonds of a macromolecule. *Note* An example is shown in Fig. 4.1.

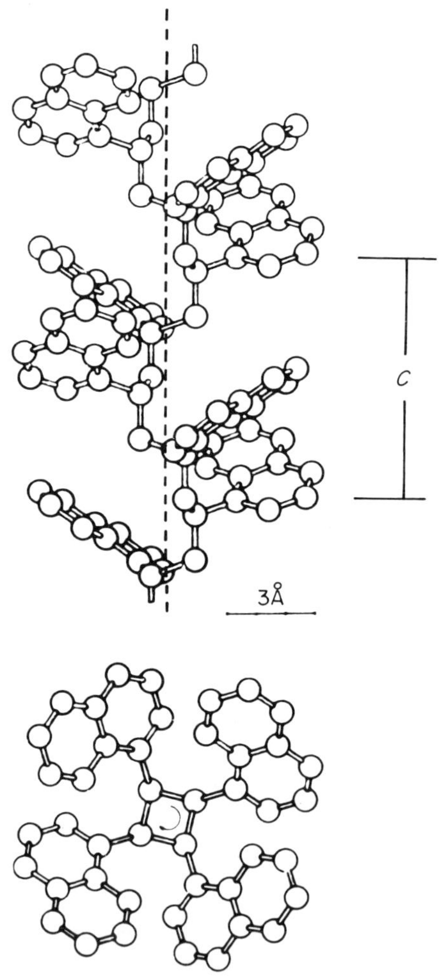

Fig. 4.1 Side view (above) and end view (below) of the macromolecule of isotactic poly(1-vinylnaphthalene) in the crystalline state. The helix symbol is s(2*4/1). The chain axis is shown by the dashed line, and *c* is the chain identity period. Hydrogen atoms are omitted. [From P. Corradini and P. Ganis. *Nuovo Cimento, Suppl.* **15**, 96 (1960)].

Term	*Definition*
2.7 helix residue	The smallest set of one or more successive configurational base units that generates the whole chain through helical symmetry.
2.8 class of helix	The number of skeletal chain atoms contained within the helix residue.
2.9 line repetition groups	The possible symmetries of arrays extending in one direction with a fixed repeating distance [3, 5, 7].

Note
Linear polymer chains in the crystalline state must belong to one of the line repetition groups (see Table 4.1 for some examples). Permitted symmetry elements are: the identity operation (symbol 1); the translation along the chain axis (symbol t); the mirror plane orthogonal to the chain axis (symbol *m*) and that containing the chain axis (symbol *d*); the glide plane containing the chain axis (symbol *c*); the inversion centre, placed on the chain axis (symbol *i*); the two-fold axis orthogonal to the chain axis (symbol 2); the helical, or screw, symmetry where the axis of the helix coincides with the chain axis. In the latter case, the symbol is s($A*M/N$), where s stands for the screw axis, A is the class of the helix, $*$ and / are separators, and M is the integral number of residues contained in N turns, corresponding to the identity period (M and N must be prime to each other) [8, 9] (see Fig. 4.1). The class index A may be dropped if deemed unnecessary, so that the helix may also be simply denoted as s(M/N) [5, 7].

Table 4.1 Chain symmetry of some crystalline polymers

Line repetition group [a]	Polymer
t1	trans-1,4-polyisoprene
s($A*M/N$)	isotactic polypropene ($M/N=3/1, A=2$)
s($A*M/N$)2	syndiotactic polypropene ($M/N=2/1, A=4$, helical modification)
t*m*	poly(heptamethylene pimelamide)
t*c*	poly(1,1-difluoroethylene) (modification 2)
t*i*	diisotactic poly[ethylene-*alt*-(*cis*-2-butene)]
s(5*2/1)*m*	poly(cyclopentene)
s(14*2/1)*d*	poly(hexamethylene adipamide)
t*cm*	syndiotactic 1,2-poly(1,3-butadiene)
s(1*2/1)*dm*	polyethylene

[a] See the note, Definition 2.9 for explanation of symbols.

2.10 structural disorder	Any deviation from the ideal three-dimensional regularity of the polymer crystal structure.

Term	*Definition*
	Note Examples of structural disorder in crystalline polymers are given in Table 4.2.

Table 4.2 Examples of structural disorder occurring in crystalline polymers

Type of structural disorder	Examples
(i) lattice distortion	as in usual crystallization (i.e. mechanical strain, lattice dislocation, impurities, etc)
(ii) chain orientation disorder	isotactic polypropene [12], isotactic polystyrene [13], poly(vinylidene fluoride)—form II [14]
(iii) configurational disorder	atactic polymers capable of crystallization: poly(vinyl alcohol) [15], poly(vinyl fluoride) [16] (see Fig. 4.2)
(iv) conformational disorder	high-temperature polymorph of trans-1,4-polybutadiene [17] (Fig. 4.3); cis-1,4-polyisoprene [17] (Fig. 4.4)
(v) macromolecular isomorphism [9, 18]	
copolymer isomorphism	poly(acetaldehyde-*co*-propionaldehyde) [19] isotactic poly[(1-butene)-*co*-(3-methyl-1-butene)] [20] isotactic poly[styrene-*co*-(*p*-fluorostyrene)] [21]
homopolymer isomorphism	mixtures of isotactic poly(4-methyl-1-pentene) and isotactic poly(4-methyl-1-hexene) [22]

2.10.1	lattice distortion	Structural disorder resulting from misalignment of the unit cells within the crystals.
2.10.2	chain-orientational disorder	Structural disorder resulting from the statistical coexistence within the crystals of identical chains with opposite orientations. *Note* A typical example is provided by the up–down statistical coexistence of anticlined chains in the same crystal structure.
2.10.3	configurational disorder	Structural disorder resulting from the statistical co-crystallization of different configurational repeating units (see Fig. 4.2).
2.10.4	conformational disorder	Structural disorder resulting from the statistical co-existence within the crystals of identical configurational units with different conformations (see Figs 4.3 and 4.4).
2.10.5	macromolecular isomorphism	Statistical co-crystallization of different constitutional repeating units, which may either belong to the same copolymer chains (copolymer isomorphism) or originate from different homopolymer chains (homopolymer isomorphism). *Note* Isomorphism is a general term: in the strict sense, the crystal structure is essentially the same throughout the range of

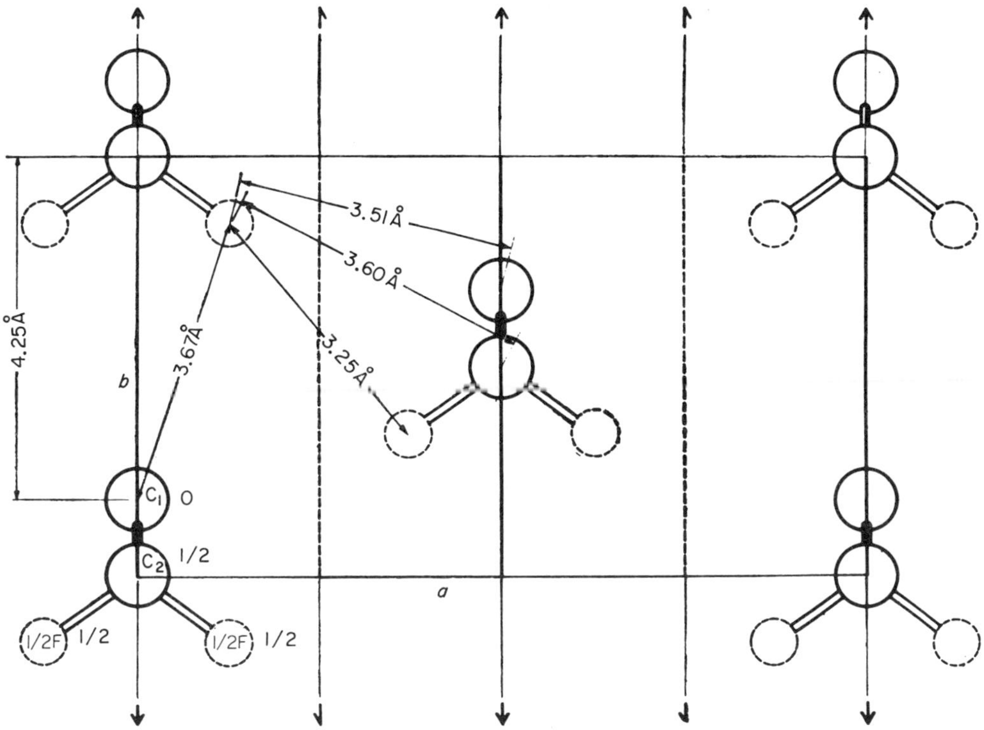

Fig. 4.2 End projection of atactic poly(vinyl fluoride) chains in the crystalline state. Broken circles show fluorine atoms with 50% probability [16].

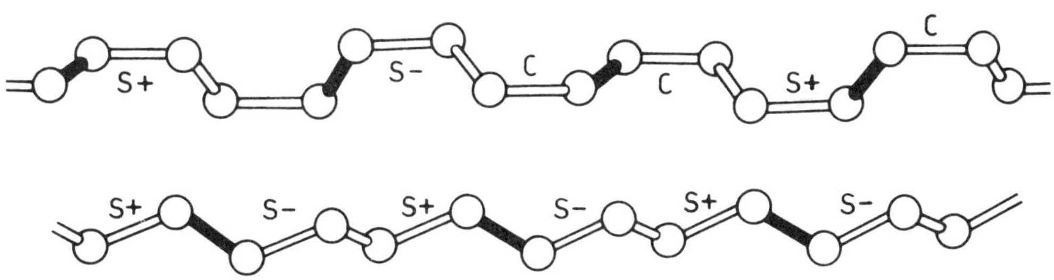

Fig. 4.3 Chain conformation of the disordered (above) and ordered (below) polymorphs of trans-1,4-polybutadiene in the crystalline state. The heavy black lines designate the double bonds and the symbols S+, S− and C the conformation [17].

Term	*Definition*
	compositions; in isodimorphism or isopolymorphism, there are two or more crystal structures, respectively, depending on composition.

3 TERMINOLOGY RELATING TO MORPHOLOGICAL ASPECTS [9, 10]

Term	*Definition*
3.1 lamellar crystal	A type of crystal with a large extension in two dimensions and a uniform thickness.

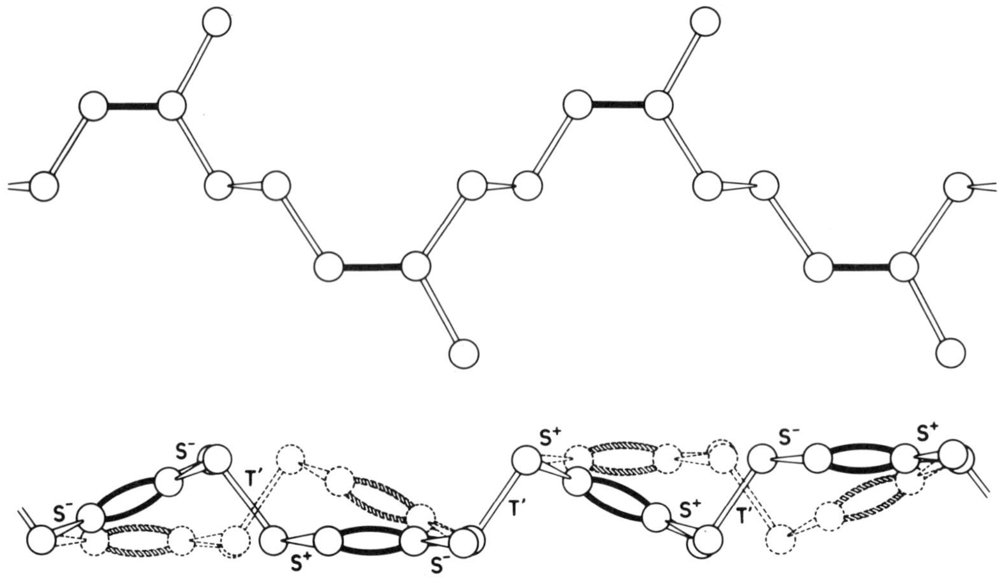

Fig. 4.4 Different possible conformations of cis-1,4-polyisoprene in the crystalline state, as viewed sideways along two orthogonal axes [17].

Term	*Definition*
	Note A lamellar crystal is usually of a thickness in the 5–50 nm range, and it may be found individually or in aggregates. The parallel-chain stems intersect the lamellar plane at an angle between 45° and 90°. The lamellae often have pyramidal shape owing to differences in the fold domains; as a result, one can deduce different fold planes and fold surfaces from the lamellar morphology.
3.2 lath crystal	A lamellar crystal prevailingly extended along one lateral dimension.
3.3 multilayer aggregate	A stack of lamellar crystals generated by spiral growth at one or more screw dislocations. *Note* The axial displacement over a full turn of the screw (Burgers vector) is usually equal to one lamellar thickness.
3.4 long spacing	The average separation between stacked lamellar crystals. *Note* The long spacing is usually measured by small-angle x-ray or neutron diffraction.
3.5 axialite	A multilayer aggregate, consisting of lamellar crystals splaying out from a common edge.

Term	*Definition*
3.6 dendrite	A crystalline morphology produced by skeletal growth, leading to a 'tree-like' appearance.
3.7 fibrous crystal	A type of crystal significantly longer in one dimension than in either of the other two. *Note* Fibrous crystals may comprise essentially extended chains parallel to the fibre axis; however, macroscopic polymer fibres containing chain-folded crystals are also known.
3.8 shish-kebab structure	A polycrystalline morphology of double habit consisting of fibrous crystals overgrown epitaxially by lamellar crystals, the stems of which are parallel to the fibre axis.
3.9 spherulite	A polycrystalline, roughly spherical morphology consisting of lath, fibrous or lamellar crystals emanating from a common centre. *Note* Space filling is achieved by branching, bending or both, of the constituent fibres or lamellae.

4 TERMINOLOGY RELATING TO MOLECULAR CONFORMATION WITHIN POLYMER CRYSTALS [9, 10]

Term	*Definition*
4.1 tie molecule	A molecule that connects at least two different crystals.
4.2 stem	A crystallized, rodlike portion of a polymer chain connected to non-rodlike portions, or chain ends, or both.
4.3 chain folding	The conformational feature in which a loop connects two parallel stems belonging to the same crystal.
4.4 fold	The loop connecting two different stems in a folded chain.
4.5 fold plane	A crystallographic plane defined by a large number of stems that are connected by chain folds.
4.6 fold surface	A surface approximately tangential to the folds.
4.7 fold domain	A portion of a polymer crystal wherein the fold planes have the same orientation. *Note* The sectors of lamellar crystals frequently represent fold domains.

Term	Definition
4.8 adjacent re-entry model	A model of crystallinity in which chain folds regularly connect adjacent stems.
4.9 switchboard model	A model of crystallinity in which the crystallized segments of a macromolecule belong to the same crystal, although the stems are connected randomly.
4.10 fringed-micelle model	A model of crystallinity in which the crystallized segments of a macromolecule belong predominantly to different crystals.
4.11 folded-chain crystal	A polymer crystal consisting predominantly of chains that traverse the crystal repeatedly by folding as they emerge at its external surfaces. *Note* The re-entry of the chain into the crystal is assumed to be adjacent or near-adjacent within the lattice.
4.12 parallel-chain crystal	A type of crystal resulting from parallel packing of stems, irrespective of the stems' directional sense.
4.13 extended-chain crystal	A polymer crystal in which the chains are in an essentially fully extended conformation.
4.14 globular-chain crystal	A type of crystal comprised of macromolecules having globular conformations. *Note* Globular-chain crystals usually occur with globular proteins.

5 TERMINOLOGY RELATING TO CRYSTALLIZATION KINETICS [10, 11]

Term	Definition
5.1 nucleation	Formation of the smallest crystalline entity, the further growth of which is thermodynamically favoured. *Note* Nucleation may be classified as primary or secondary. Primary nucleation can be homogeneous or heterogeneous; if heterogeneous nucleation is initiated by entities having the same composition as the crystallizing polymer, it is called self-nucleation. Secondary nucleation is also known as surface nucleation.
5.2 molecular nucleation	Initial crystallization of a small portion of a macromolecule, after which further crystallization is thermodynamically favoured.

Term	*Definition*
	Note Molecular nucleation may give rise to a new crystal or increase the size of a pre-existing one.
5.3 Avrami equation	An equation describing crystallization kinetics. *Notes* 1. The Avrami equation has the form $1-\phi_c=\exp(-Kt^n)$ (6) where ϕ_c is the crystalline volume fraction developed at time t and constant temperature, and K and n are suitable parameters. 2. K in equation (6) is temperature-dependent. 3. According to the original theory, n in equation (6) should be an integer from 1 to 4, the value of which should depend only on the type of the statistical model; however, it has become customary to regard it as an adjustable parameter that may be non-integral.
5.4 primary crystallization	The first stage of crystallization, considered to be ended when most of the spherulite surfaces impinge on each other. *Note* In isothermal crystallization, primary crystallization is often described by the Avrami equation.
5.5 secondary crystallization	Crystallization occurring after primary crystallization, usually proceeding at a lower rate.
5.6 reorganization	The molecular process by which (i) amorphous or poorly ordered regions of a polymer specimen become incorporated into crystals, or (ii) a change to a more stable crystal structure takes place, or (iii) defects within the crystals decrease. *Notes* 1. Secondary crystallization may be involved in the reorganization process. 2. Reorganization may result from annealing. 3. (i) and (iii) may also be called crystal perfection.
5.7 recrystallization	Reorganization proceeding through partial melting. *Note* Recrystallization is likely to result in an increase in the degree of crystallinity, or crystal perfection, or both.
5.8 segregation	The rejection of a fraction of macromolecules, or of impurities, or both, from growing crystals.

Note
The rejected macromolecules are usually those of insufficient relative molecular mass, or differing in constitution or configuration (e.g. branching, tacticity, etc).

REFERENCES

1 IUPAC. Basic definitions of terms relating to polymers 1974. *Pure Appl. Chem.* **40**, 477–491 (1974). Reprinted as Chapter 1, this volume.

2 IUPAC. Definitions of terms relating to individual macromolecules, their assemblies, and dilute polymer solutions (recommendations 1988). *Pure Appl. Chem.* **61**, 211–241 (1989). Reprinted as Chapter 3, this volume.

3 The International Union of Crystallography. *International Tables for X-ray Crystallography.* Kynoch Press, Birmingham, UK, Vol. **I** (1969).

4 H. Tadokoro. *Structure of Crystalline Polymers.* Wiley–Interscience, New York (1979).

5 IUPAC. Stereochemical definitions and notations relating to polymers (recommendations 1980). *Pure Appl. Chem.* **53**, 733–752 (1981). Reprinted as Chapter 2, this volume.

6a C.W. Bunn. The stereochemistry of chain polymers. *Proc. R. Soc. London, A* **180**, 67 (1942).

6b M.L. Huggins. Comparison of the structures of stretched linear polymers. *J. Chem. Phys.* **13**, 37 (1945).

6c L. Pauling, R.B. Corey and H.R. Branson. The structure of proteins: two hydrogen-bonded helical configurations of the polypeptide chain. *Proc. Natl. Acad. Sci. USA* **37**, 205 (1951).

6d G. Natta and P. Corradini. Structure of crystalline polyhydrocarbons. *Nuovo Cimento, Suppl.* **15**, 9 (1960).

7 P. Corradini. Chain conformation and crystallinity. In *Stereochemistry of Macromolecules* (Ed. A. Ketley). Marcel Dekker, New York, Vol. **III**, 1–60 (1968).

8 R.L. Miller. Crystallographic data for various polymers. In *Polymer Handbook*, 2nd Edn (Eds. J. Brandrup and E.H. Immergut). Wiley–Interscience, New York, III-1 (1975).

9 B. Wunderlich. *Macromolecular Physics.* Academic Press, New York, Vol. **1** (1973).

10 P.H. Geil. *Polymer Single Crystals.* Interscience, New York (1963).

11 B. Wunderlich. *Macromolecular Physics.* Academic Press, New York, Vol. **2** (1976).

12 G. Natta and P. Corradini. Structure and properties of isotactic polypropylene. *Nuovo Cimento, Suppl.* **15**, 40 (1960).

13 G. Natta, P. Corradini and I.W. Bassi. Crystal structure of isotactic polystyrene. *Nuovo Cimento, Suppl.* **15**, 68 (1960).

14 Y. Takahashi and H. Tadokoro. Short-range order in form II of poly(vinylidene fluoride): antiphase domain structures. *Macromolecules* **16**, 1880 (1983).

15 C.W. Bunn and H.S. Peiser. Mixed crystal formation in high polymers. *Nature* **159**, 161 (1947).

16 G. Natta, I.W. Bassi and G. Allegra. Struttura crystallina del polivinilfloruro atattico. *Atti Accad. Naz. Lincei Rend., Cl. Sci. Fis., Mat. Nat.* **31**, 350 (1961).

17 P. Corradini. Observation of different conformations of a macromolecule in the crystalline state. *J. Polym. Sci., Polym. Symp. Ed.* **51**, 1 (1975).

18 G. Allegra and I.W. Bassi. Isomorphism in synthetic macromolecular systems. *Adv. Polym. Sci.* **6**, 549 (1969).

19 A. Tanaka, Y. Hozumi, K. Hatada, S. Endo and R. Fujishige. Isomorphism phenomena in polyaldehydes. *J. Polym. Sci., Polym. Lett. Ed.* **2**, 181 (1964).

20 A. Turner Jones. Crystalline phases in copolymers of butene and 3-methylbutene. *J. Polym. Sci., Polym. Lett. Ed.* **3**, 591 (1965).

21 G. Natta, P. Corradini, D. Sianesi and D. Morero. Isomorphism phenomena in macromolecules. *J. Polym. Sci.* **51**, 527 (1961).

22 G. Natta, G. Allegra, I.W. Bassi, C. Carlini, E. Chiellini and G. Montagnoli. Isomorphism phenomena in isotactic poly(4-methyl-substituted α-olefins) and in isotactic poly(alkyl vinyl ethers). *Macromolecules* **2**, 311 (1969).

ALPHABETICAL INDEX OF TERMS

Nomenclature

5: Nomenclature of Regular Single-Strand Organic Polymers (1975)

PREAMBLE

In 1952, the Subcommission on Nomenclature of the IUPAC Commission on Macromolecules published a report [1] on the nomenclature of macromolecules that included a method for the systematic naming of linear organic polymers on the basis of structure. A later report [2] dealing with steric regularity utilized this system of nomenclature. When the first report was issued, the skeletal rules were adequate for most needs; indeed, most polymers could at that time be reasonably named on the basis of the substance used in producing the polymer. In the intervening years, however, the rapid growth of the polymer field has dictated a need for modification and expansion of the earlier rules. This report presents an updating of those rules. Necessarily, a great many changes in detail were required, since it is desirable that organic polymer nomenclature adhere as much as possible to the definitive rules for the Nomenclature of Organic Chemistry [3].

These rules are designed to name, uniquely and unambiguously, the structures of regular single-strand organic polymers whose repeating structures can be written within the framework of ordinary chemical principles; stereochemistry is not considered in this report. As with organic nomenclature, this nomenclature describes chemical structures rather than substances. It is realized that polymeric substances ordinarily include many structures, and that a complete description of even a single polymer molecule would include an itemization of terminal groups, branching, random impurities, degree of steric regularity, chain imperfections, etc. Nevertheless, it is useful to think of a substance as being represented by a single structure that may itself be hypothetical. To the extent that the polymer structure can be portrayed as a chain of regularly repeating structural or constitutional repeating units (the terms are synonymous), the structure can be named by these rules; in addition, provision has been made for including end groups in the name.

In this report, the fundamental principles and the basic rules of the structure-based nomenclature are given first, accompanied by detailed extensions and applications. An Appendix is included containing a limiting list of acceptable source-based names, along with the corresponding structure-based names, of common polymers. The Commission sees no objection to the continued use of such source-based names where these names are clear and unambiguous, but prefers the use of the structure-based nomenclature detailed in these rules.

FUNDAMENTAL PRINCIPLES

This nomenclature system rests upon the selection of a preferred *constitutional repeating unit* [4] (CRU) of which the polymer is a multiple; the name of the polymer is simply the name of this

Those served on the Commission during the preparation of this document were K.L. Loening (*Chairman*), R.B. Fox (*Secretary*), N.M. Bikales, P. Corradini, L.C. Cross, A.D. Jenkins, N.A. Platé, W. Ring, P. Sigwalt, G.J. Smets and T. Tsuruta.
Reprinted from *Pure Appl. Chem.* **48**, 373–385 (1976).

repeating unit prefixed by *poly*. The unit itself is named wherever possible according to the definitive rules for the Nomenclature of Organic Chemistry [3]. For regular single-strand polymers, this unit usually is a bivalent group.

In using this nomenclature, the steps to be followed in sequence are (1) *identify* the CRU; (2) *orient* the CRU; and (3) *name* the CRU. Identification and orientation must always precede the selection of the name of the polymer.

Identification of the constitutional repeating unit

There are many ways to write the CRU for most polymer structures. In simple cases, these units are readily identified. For example, in

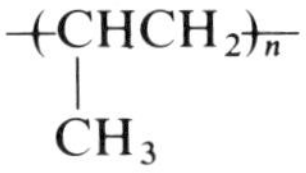

the CRUs are

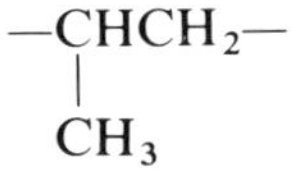

and

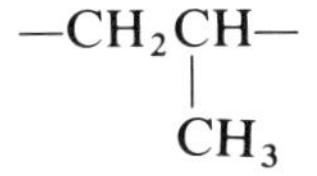

In more complex cases, it is often necessary to draw a large segment of the chain and from it choose all of the possible CRUs. For example, in the polymer

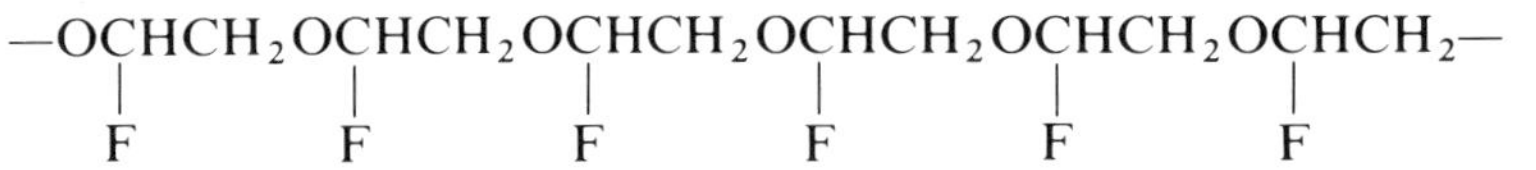

the CRUs are

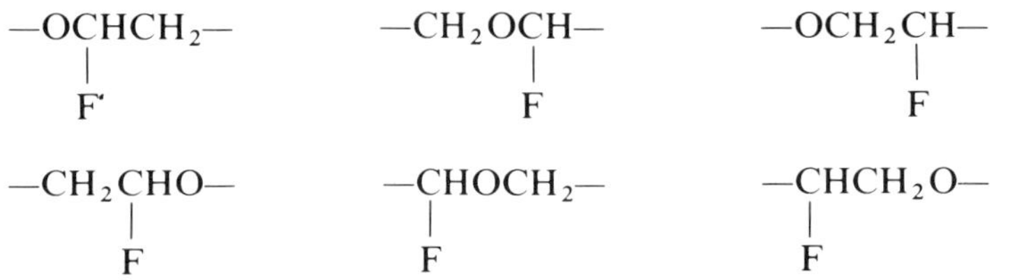

To allow construction of a unique name, a single CRU must be selected. The following rules have been designed to specify both *seniority* among subunits, i.e. the point at which to begin writing the CRU, and the *direction* along the chain in which to continue to the end of the CRU. The preferred constitutional repeating unit will be one beginning with the subunit of highest seniority (see Rule 2). From this subunit, one proceeds toward the subunit next in seniority. In the preceding example, the subunit of highest seniority is an oxygen atom and the subunit next in seniority is a substituted $—CH_2CH_2—$ unit. The parent CRU will therefore be either $—OCH_2CH_2—$ or $—CH_2CH_2O—$. Further choice in this case is based on the lowest locant for substitution, so that the CRU is

rather than

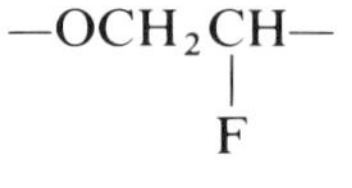

or

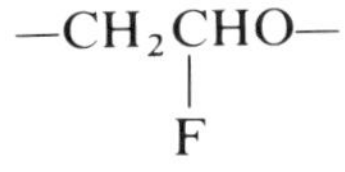

rather than

Orientation of the constitutional repeating unit
The CRU is written to read from left to right. In the above example, the preferred CRU is therefore

$$-\mathrm{O\underset{\displaystyle F}{\underset{|}{C}}HCH_2}-$$

Naming the constitutional repeating unit
The name of the CRU is formed by citing, in order, the names of the largest subunits within the CRU (Rule 1.2.1). In the example, the oxygen atom is called oxy and the $-CH_2CH_2-$ (preferred to $-CH_2-$ because it is larger and can be named as a unit) is called ethylene; the latter unit substituted with one fluorine atom is called 1-fluoroethylene. The CRU in question is therefore named oxy(1-fluoroethylene), and the corresponding polymer is

$$\mathrm{-\!\!\!(O\underset{\displaystyle F}{\underset{|}{C}}HCH_2)\!\!\!-_n} \qquad \text{poly[oxy(1-fluoroethylene)]}$$

The rules which follow are essentially directions for the selection of the CRU in a given polymer.

RULE 1 THE CONSTITUTIONAL REPEATING UNIT

Regular single-strand polymer chains can usually be represented as multiples of a bivalent repeating unit which can itself be named. The name of the polymer is *poly(constitutional repeating unit)*. In those cases in which a choice is possible between a bivalent and a higher-valent CRU, the number of free valences* is minimized only after all other orders of seniority have been observed (see also Rule 2.1.2).

$-CH{=}CH-$ is preferred to $=CH-CH=$

but

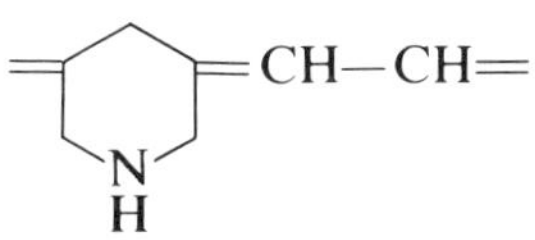 is preferred to 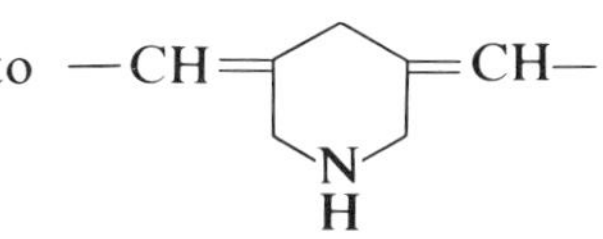

1.1 The generic name

Linear polymers of unspecified chain length will be named by prefixing *poly* to the name, placed in parentheses or brackets, of the structural repeating unit of the polymer, i.e. the smallest unit of which the polymer is a multiple. If the name of the repeating unit is 'ABC', the corresponding polymer name is

$-\!\!\!(\mathrm{ABC})\!\!\!-_n$ poly(ABC)

Where it is desired to specify chain length, the appropriate Greek prefix (deca, docosa, etc) may be used in place of *poly*. For a single-strand polymer, the CRU is named within the restriction of directional citation by the IUPAC organic nomenclature rules [3].

*In this document 'free valence' means classical free valence.

1.2 Simple constitutional repeating units

1.2.1

The CRU may contain one or more subunits. Among the possible subunits or combinations of adjacent subunits, the largest possible group, based on main chain atoms and rings only, is to be named (see also Rule 2.1.1). When the largest group includes the entire CRU, its name, prefixed by *poly* is the name of the polymer.

$\left(CH_2 \right)_n$ poly(methylene)

$\left(OCH_2CH_2 \right)_n$ poly(oxyethylene)

The name of a CRU or any subunit has no relationship to the manner in which the unit was prepared; the name is simply that of the largest identifiable unit and any locants for unsaturation, substituents, etc are dictated by the *structure* of the unit.

$\left(CH{=}CHCH_2CH_2 \right)_n$

poly(1-butenylene)

(not poly(2-butenylene), which gives a higher locant to the double bond, nor poly(vinyleneethylene), which identifies less than the largest unit in the CRU)

1.2.2

Identification of the preferred CRU rests on (i) the kinds of atoms or rings in the main chain; or (ii) on the location of substituents when there is only one kind of main chain atom or ring. Orientation of the CRU in case (i) is determined by the rules of seniority given in Rule 2; in case (ii), lowest locants (except when fixed numbering applies; see Rule 1.2.4) are given to substituents in alphabetical order (Rule 2.4.2).

$\left(\underset{\displaystyle D}{\underset{|}{C}}HCH_2CH_2 \right)_n$ poly(trimethylene-1-*d*)

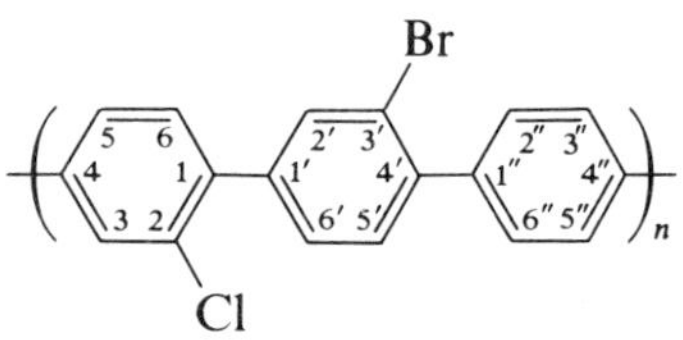

poly(3′-bromo-2-chloro-*p*-terphenyl-4,4″-ylene)

After the CRU and its orientation, reading left to right, have been established, the CRU or its constituent subunits are named to include as many as possible, in order, of (i) the main chain atoms or rings; and (ii) the substituents within a single name (see also Rule 3.1).

$\left(\underset{\displaystyle CH_3}{\underset{|}{C}}H \right)_n$ poly(ethylidene)
(not poly(methylmethylene))

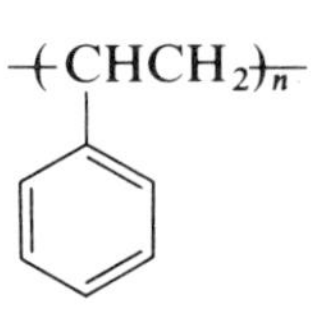

poly(1-phenylethylene)
(not poly(benzylidenemethylene) or poly(1-phenyldimethylene)

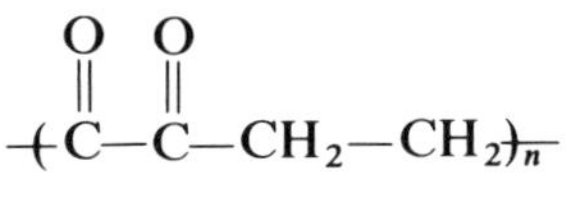

poly(1,2-dioxotetramethylene)
(not poly(succinyl), since substituent positions 1,2 are preferred to 1,4, and identification and orientation of the CRU precede formation of the name)

poly(oxysuccinyl)
(not poly[oxy(1,4-dioxotetramethylene)], since succinyl is an approved name [3])

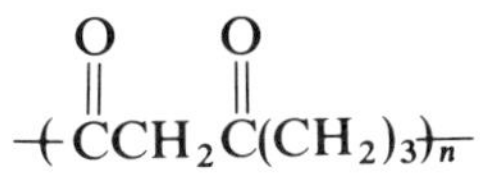

poly(1,3-dioxohexamethylene)
(not poly(malonyltrimethylene) because the six-carbon chain is the largest unit that can itself be named)

$\text{-(}CH{=}CH\text{)}_n\text{-}$ poly(vinylene)
(not $\text{=(}CH{-}CH\text{)}_n\text{=}$, poly(ethanediylidene))

1.2.3
If, after identification and orientation, the CRU is found to contain one or more acyclic bivalent groups having more than two hetero atoms in the main chain, these groups may often be advantageously named by replacement nomenclature [3]. The main chain of the group is named and numbered as though the entire chain were an acyclic hydrocarbon and the hetero atoms are named by means of prefixes 'aza', 'oxa', etc with locants to fix their positions.

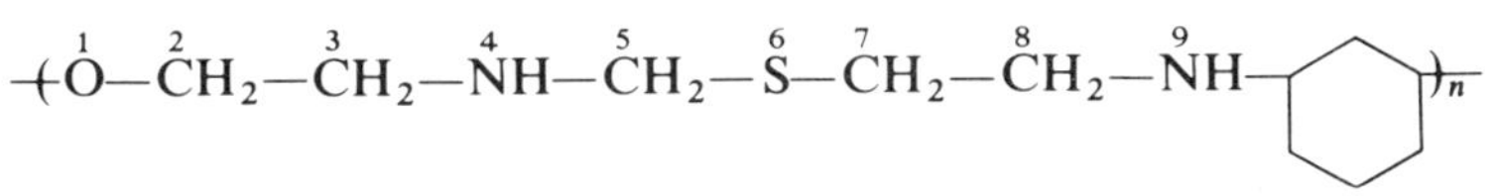

Replacement name: poly(1-oxa-6-thia-4,9-diazanonamethylene-1,3-cyclohexylene)
Systematic name: poly(oxyethyleneiminomethylenethioethyleneimino-1,3-cyclohexylene)

See Rules 2.1.4 and 2.3.2 for other examples of the use of replacement nomenclature.

1.2.4
Groups having fixed numbering retain that numbering in naming the CRU (see also Rules 2.2.2 and 2.4.1).

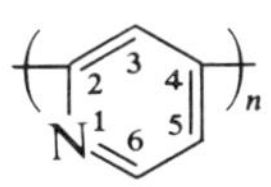

poly(2,4-pyridinediyl)

For most acyclic and monocarbocyclic groups, preference in lowest numbers is given to the carbon atoms having the free valences. In other families of compounds, notably the polycyclic hydrocarbons, bridged hydrocarbons, spiro hydrocarbons and heterocyclic ring systems, numbering is fixed for the ring system. Free valences in groups are numbered as low as possible, consistent with the fixed numbering. Since direction through the bivalent group is a requisite parameter in naming polymers, the same fixed numbering is retained for either direction of progress through the group in generating the polymer name.

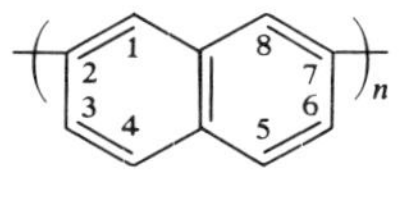

poly(2,7-naphthylene)
(not poly(7,2-naphthylene) or poly(3,6-naphthylene))

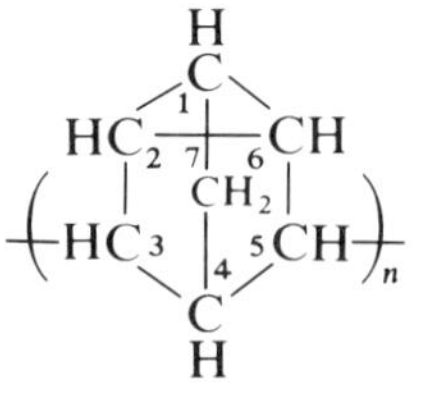

poly(tricyclo[2.2.1.0^{2,6}]hept-3,5-ylene)

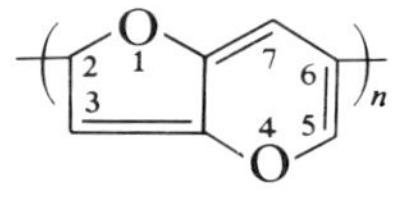

poly(2*H*-furo[3,2-*b*]pyran-2,6-diyl)

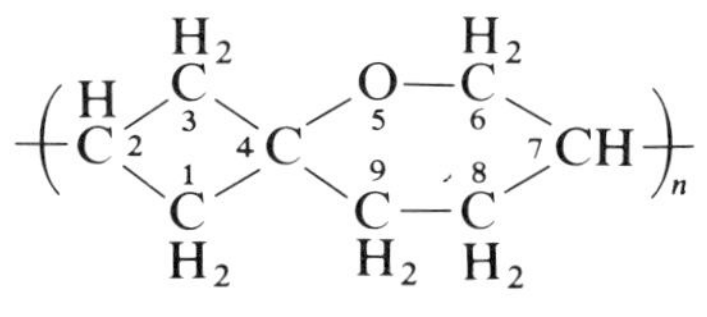

poly(5-oxaspiro[3.5]-non-2,7-ylene)

RULE 2 CONSTITUTIONAL REPEATING UNITS HAVING TWO OR MORE SUBUNITS

Many regular single-strand polymers can be represented as multiples of repeating units, such as —ABC—, which consist of a series of smaller subunits, —A—, —B—, and —C—. The prototype name of the polymer is *poly*(*ABC*), where (ABC) stands for the names of A, B and C, taken in that order. This rule is concerned with the seniority of subunits in identifying the preferred CRU for a given polymer structure.

2.1 Seniority of subunits and direction of citation

2.1.1

Polymers having CRUs containing two or more subunits are named with the prefix *poly* followed in parentheses or brackets by the names of the largest possible subunits cited in order from left to right as they appear in the CRU. The CRU is written from left to right beginning with the subunit of highest seniority and proceeding in a direction defined by the shorter path to the subunit next in seniority.

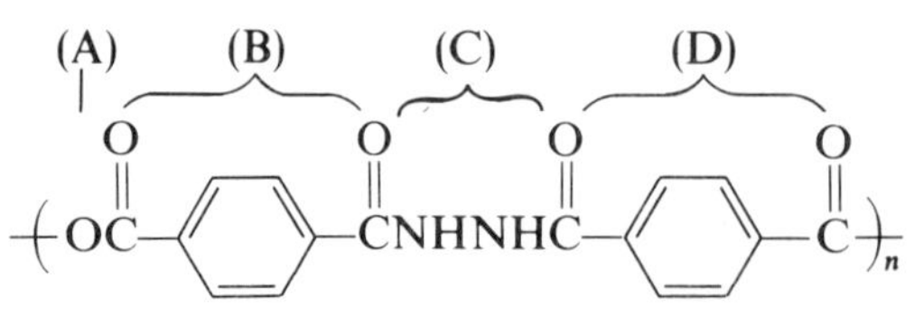

poly(oxyterephthaloylhydrazoterephthaloyl)
(not poly(oxycarbonyl-1,4-phenylenebicarbamoyl-1,4-phenylenecarbonyl))

2.1.2

The CRU in a regular single-strand polymer should usually be a bivalent group. The starting point for the unit is at the subunit of highest seniority, and citation will be in the direction of the shorter path toward that subunit or subunit combinations of highest seniority.

For citation of the first subunit, the order of seniority among the types of bivalent groups is (1) *heterocyclic rings* (see Rule 2.2); followed by (2) *chains containing hetero atoms* (see Rule 2.3); (3) *carbocyclic rings* (see Rule 2.4); and (4) *chains containing only carbon*, in that order. This order is unaffected by the presence of rings, atoms or groups that are not part of the main chain, even though such substituents could be expressed as part of a trivial name for a bivalent group.

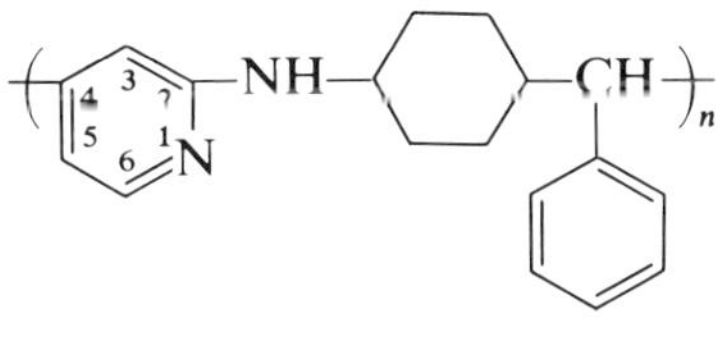

poly(4,2-pyridinediylimino-1,4-cyclohexylenebenzylidene)

2.1.3

Choice of direction along the main chain of the CRU is determined by the shorter path, counting ring and chain atoms individually, from the subunit of highest seniority to the subunit next in seniority.

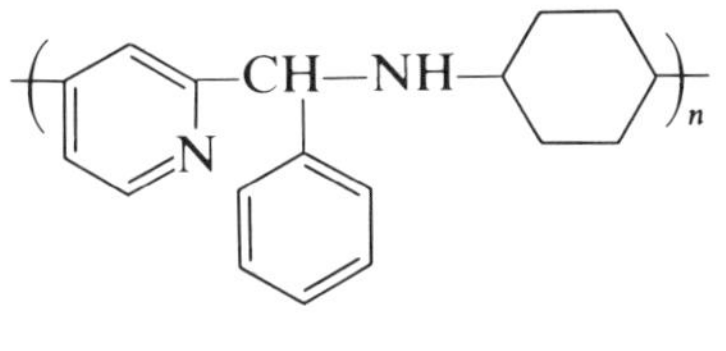

poly(4,2-pyridinediylbenzylideneimino-1,4-cyclohexylene)

The possible paths between subunits of first and second seniority necessarily involve subunits of lesser seniority. Except in cases where two paths are of equal length (see Rule 2.1.4), the number rather than the nature of the atoms involved is the determining factor.

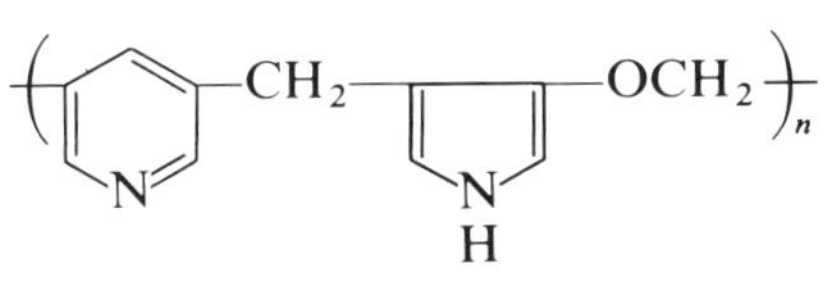

poly(3,5-pyridinediylmethylenepyrrole-3,4-diyloxymethylene)
(not poly(3,5-pyridinediylmethyleneoxypyrrole-3,4-diylmethylene), in which the longer $—CH_2O—$ path between rings is followed)

Where a ring constitutes all or part of a path, the shortest continuous chain of atoms in the ring is selected.

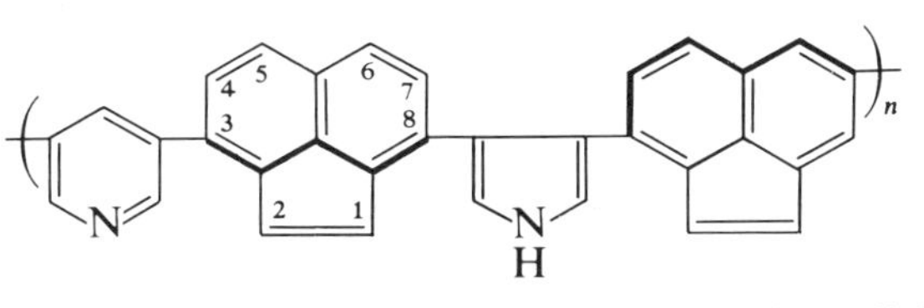

poly(3,5-pyridinediyl-3,8-acenaphthylenylenepyrrole-3,4-diyl-3,7-acenaphthylenylene)
(heavy line denotes path followed)

2.1.4

When the choice of path determining direction of citation involves paths of equal length to subunits of equal seniority in the normal order of precedence, the choice of path depends upon the kind of subunits in the paths themselves. This condition applies to chains typified by the following generalized structures, where A, B and C are subunits in that order of decreasing seniority, separated by paths of differing lengths x and y which contain units of lower seniority than C:

—C—y—B—x—A—x—B—y—C—

—B—y—A—x—A—y—B—

—B—y—A—x—A—x—A—y—B—

The choice of direction is from a subunit A to the nearest part of a path x having highest seniority, or, if two paths x are identical in every respect, to the nearest part of path y having highest seniority etc, until some point of difference is encountered (see also Rules 2.2.4 and 2.3.2).

Examples of direction of citation based on the constituent parts of paths of equal length are:

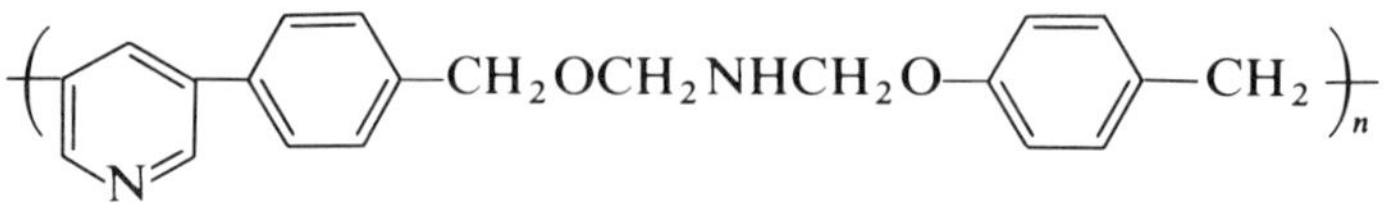

poly(3,5-pyridinediyl-1,4-phenylenemethyleneoxymethyleneiminomethyleneoxy-1,4-phenylenemethylene)
(choice of path from heterocyclic ring to O determined by position of phenylene)

$\text{-(}OCH_2OCH_2NHCH_2CH_2SCH_2NHCH_2CH_2\text{)}_n$

poly(oxymethyleneoxymethyleneiminoethylenethiomethyleneiminoethylene) or poly(1,3-dioxa-8-thia-5,10-diazadodecamethylene)
(choice of path from O to S determined by position of NH)

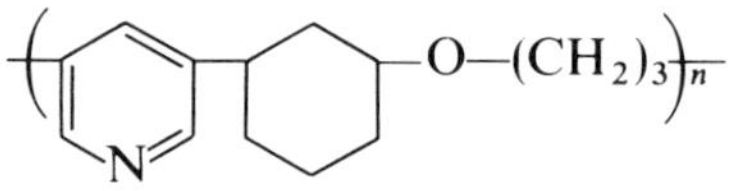

poly(3,5-pyridinediyl-1,3-cyclohexyleneoxytrimethylene)
(a portion of a cyclic structure is senior to carbon chain of equal length)

Where substituents control the choice of CRU, the order of seniority is that given in Rule 2.4.2.

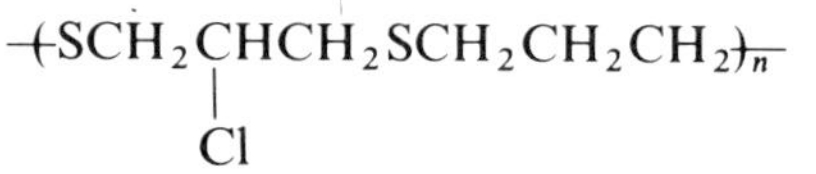 poly[thio(2-chlorotrimethylene)thiotrimethylene]

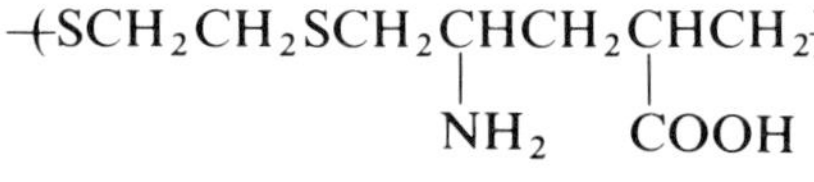

poly[thioethylenethio(2-amino-4-carboxypentamethylene)]
(direction determined by alphabetical order)

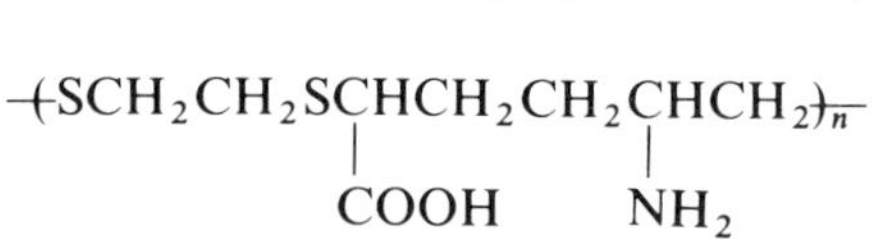

poly[thioethylenethio(4-amino-1-carboxypentamethylene)]
(direction determined by lowest locants takes precedence over alphabetical order)

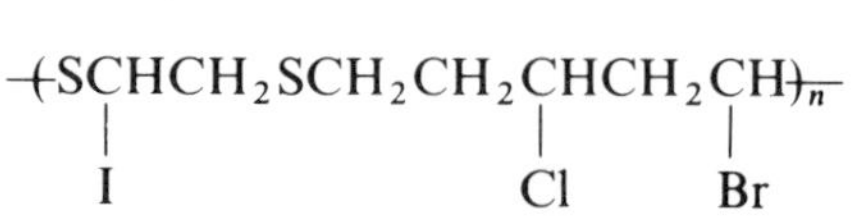

poly[thio(1-iodoethylene)thio(5-bromo-3-chloropentamethylene)]
(direction determined by the lower locant in the first cited subunit after beginning the CRU)

2.2 Heterocyclic rings

2.2.1

A CRU having two or more subunits that include a heterocyclic ring system in the main chain is named by citing first the heterocyclic ring group of highest seniority and proceeding by the shorter path in descending order of preference to (i) another of the same heterocyclic ring (see Rule 2.2.4)

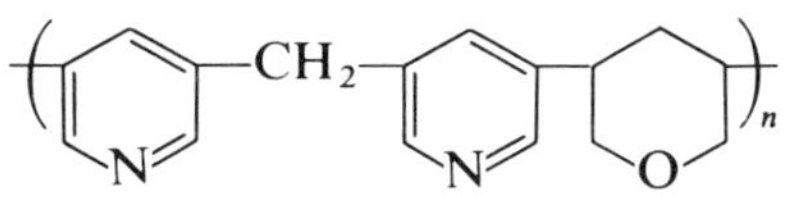

poly[3,5-pyridinediylmethylene-3,5-pyridinediyl(tetrahydro-2*H*-pyran-3,5-diyl)]

(ii) the heterocyclic ring next in seniority (see Rule 2.2.3)

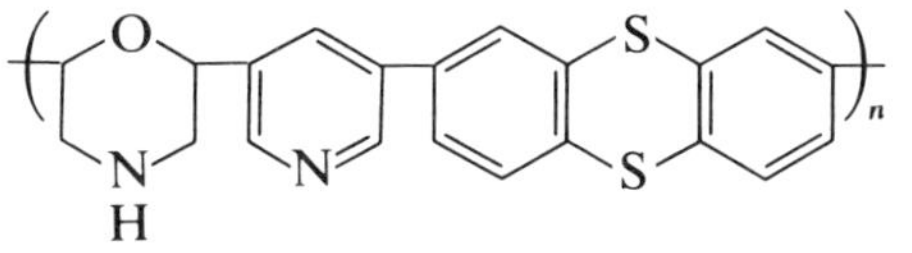

poly(2,6-morpholinediyl-3,5-pyridinediyl-2,8-thianthrenediyl)

(iii) the senior acyclic group containing a hetero atom in the main chain (see Rule 2.3.1)

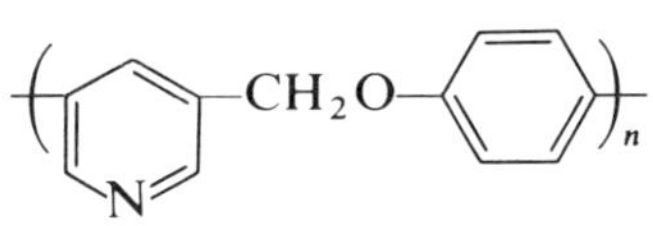

poly(3,5-pyridinediylmethyleneoxy-1,4-phenylene)

(iv) the senior carbocyclic ring system (see Rule 2.4.1)

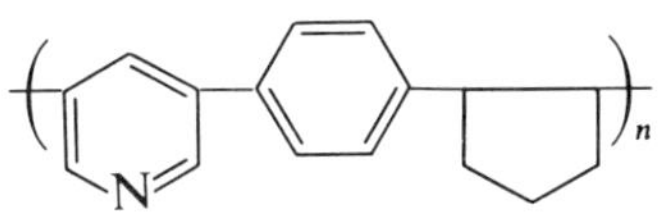

poly(3,5-pyridinediyl-1,4-phenylene-1,2-cyclopentylene)

and (v) the senior acyclic group containing only carbon in the main chain (see Rule 2.4.2)

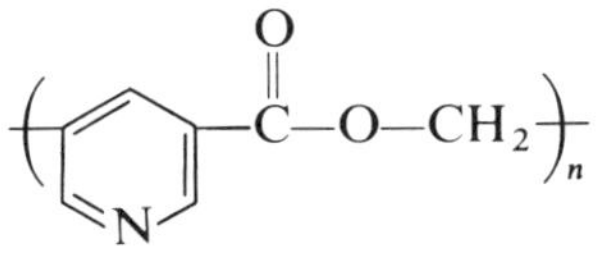

poly(3,5-pyridinediylcarbonyloxymethylene)

2.2.2

Consistent with the fixed numbering of heterocyclic rings, the points of attachment to the main chain of the CRU should have the lowest permissible locants.

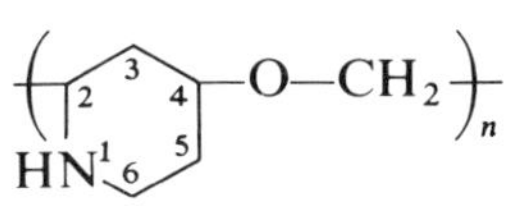

poly(2,4-piperidinediyloxymethylene)

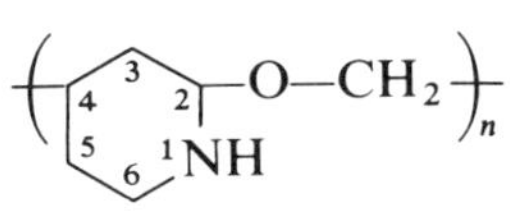

poly(4,2-piperidinediyloxymethylene)

Where there is a choice, the point of attachment at the left side of the ring should have the lowest permissible number.

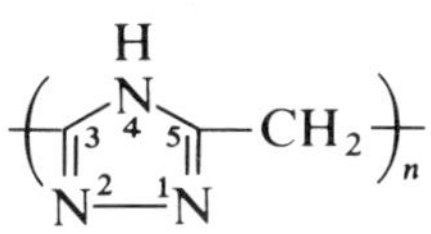

poly(4*H*-1,2,4-triazole-3,5-diylmethylene)

2.2.3

Among heterocyclic ring systems, the descending order of seniority is (i) a ring system with nitrogen in the ring; (ii) a ring system containing a hetero atom other than nitrogen as high as possible in the order given in Rule 2.3.1; (iii) a ring system containing the greatest number of rings; (iv) a ring system having the largest individual ring; (v) a ring system having the largest number of hetero atoms; (vi) a ring system containing the greatest variety of hetero atoms; and (vii) the ring system having the greatest number of hetero atoms highest in the order given in Rule 2.3.1. This order is that followed in Table B-I of the IUPAC organic nomenclature rules [3].

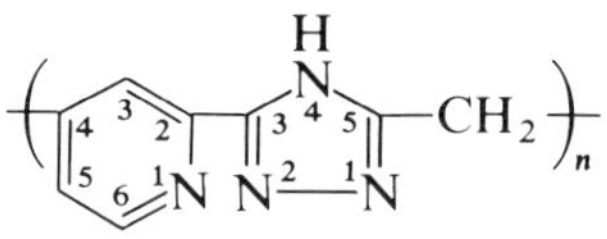

poly(4,2-pyridinediyl-4*H*-1,2,4-triazole-3,5-diylmethylene)

Further examples of the application of seniority in heterocyclic ring systems are

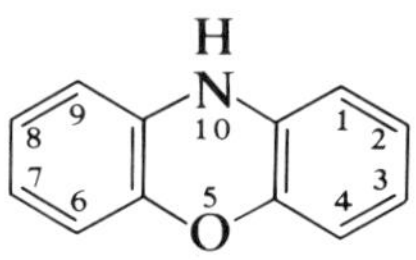

phenoxazine

>

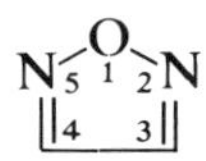

furazan

>

thiazole

>

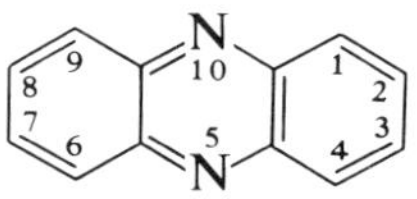

phenazine

>

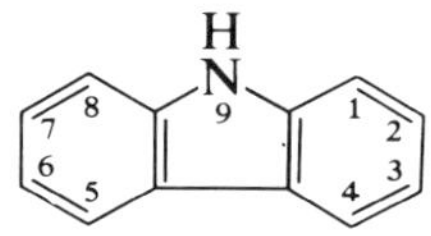

carbazole

>

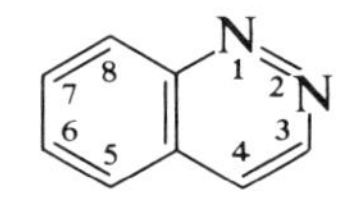

cinnoline

>

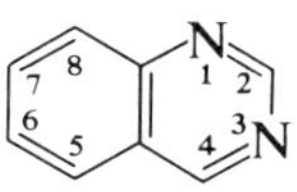

quinazoline

>

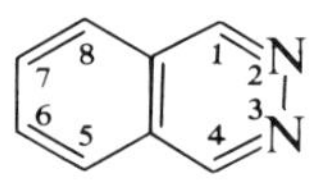

phthalazine

>

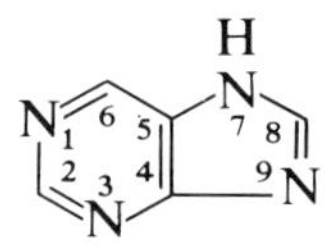

purine

>

pyrimidine

>

pyridine

>

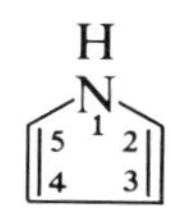

pyrrole

>

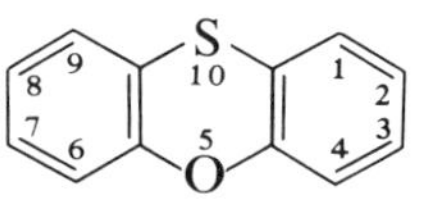

phenoxathiin

>

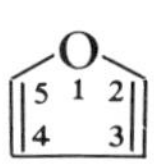

furan

>

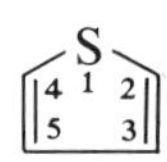

thiophene

When two heterocyclic subunits differ only in degree of unsaturation, the senior subunit is that having the least hydrogenated ring system.

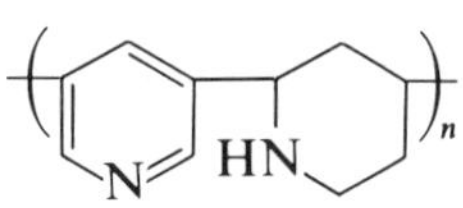

poly(3,5-pyridinediyl-2,4-piperidinediyl)

Among assemblies of identical heterocyclic rings, the ring of highest seniority is that having lowest numbers for the points of attachment between the rings within the assembly consistent with the fixed numbering of the parent ring.

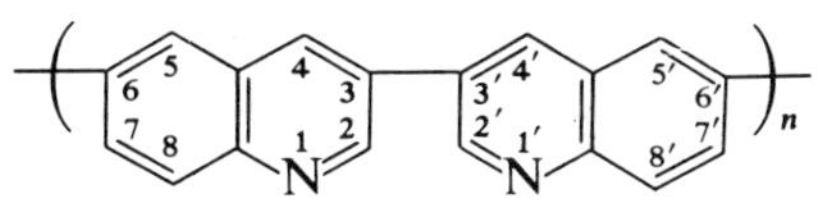

poly([3,3′-biquinoline]-6,6′-diyl)
(not poly([6,6′-biquinoline]-3,3′-diyl))

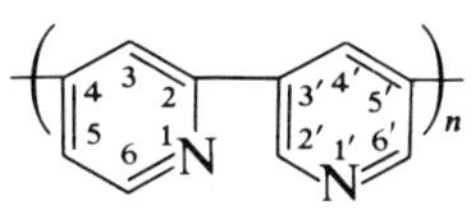

poly([2,3′-bipyridine]-4,5′-diyl)
(not poly([2′,3-bipyridine]-4′,5-diyl) or poly([3′,4-bipyridine]-2,5′-diyl))

Further choice is based on the number and kind of substituting groups (see Rule 2.4.2).

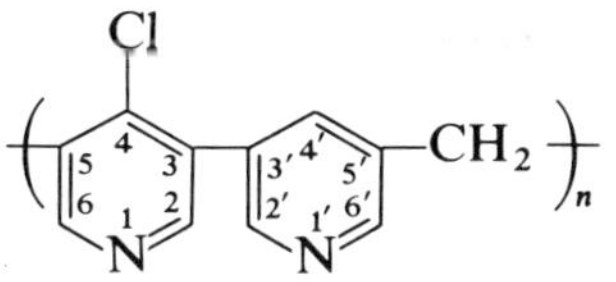

poly[(4-chloro-[3,3′-bipyridine]-5,5′-diyl)methylene]

2.2.4

When the CRU contains two identical rings of highest seniority or more than two such rings of highest seniority separated by identical paths, the direction of citation is determined by the shorter path to the subunit of second seniority. Further choice is based on the shorter path from that subunit to the subunit of third seniority, etc, as indicated in the order of seniority in Rule 2.2.1.

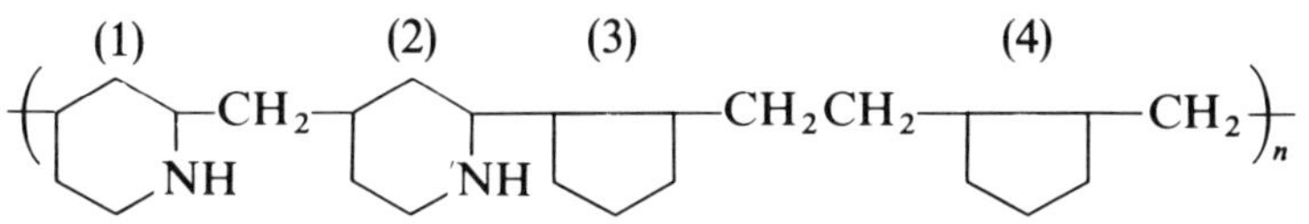

poly(4,2-piperidinediylmethylene-4,2-piperidinediyl-1,2-cyclopentyleneethylene-1,2-cyclopentylenemethylene)
(numbers above rings show order of preference in formation of name)

2.3 Hetero atoms in chains

2.3.1

Complex CRUs in which the senior subunit is a hetero atom or an acyclic chain with a hetero atom in the main chain are named by citing first the hetero atom of highest seniority and proceeding by the shorter path in descending order of seniority to (i) another hetero atom of the same kind; (ii) the hetero atom next in seniority; (iii) the senior carbocyclic ring system (see Rule 2.4.1); and (iv) the senior acyclic group containing only carbon in the main chain (see Rule 2.4.2). For the most common hetero atoms the descending order of seniority is O, S, Se, Te, N, P, As, Sb, Bi, Si, Ge, Sn, Pb, B and Hg; other hetero atoms may be placed within this order as indicated by their positions in the periodic table.

$$-\!\left(NH\overset{\overset{\displaystyle O}{\|}}{C}CH_2SiH_2(CH_2)_3\right)_{\!n}\!-$$

poly[imino(1-oxoethylene)silylenetrimethylene]

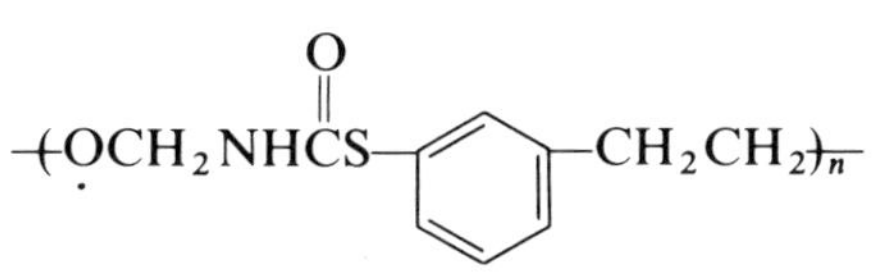

poly(oxymethyleneiminocarbonylthio-1,3-phenyleneethylene)

$$-\!\left(ONHCH_2NHNHCH_2\right)_{\!n}\!-$$

poly(oxyiminomethylenehydrazomethylene)

Parentheses must be used in some cases to prevent ambiguity.

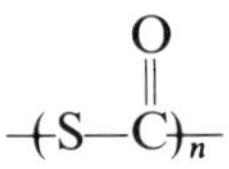

poly[thio(carbonyl)]
('carbonyl' is enclosed in parentheses to differentiate the structure from 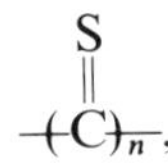 poly(thiocarbonyl))

The direction of bonding in unsymmetrical single-atom radicals (e.g. =N— or —N= for nitrilo) is indicated by the endings of the names of the adjacent subunits in the CRU.

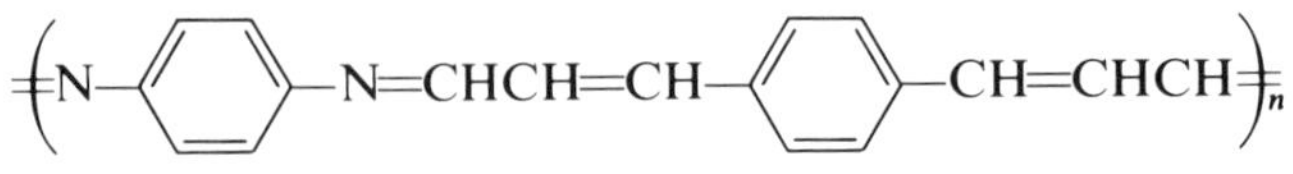

poly(nitrilo-1,4-phenylenenitrilo-2-propen-3-yl-1-ylidene-1,4-phenylene-1-propen-1-yl-3-ylidene)

 poly(oxycarbonylnitrilo-1,3-propanediylidenenitrilocarbonyl)

Direction in a group such as azoxy (—N(→O)=N— or —N=N(→O)—) is indicated by the prefixes ONN or NNO, respectively, in that order of seniority.

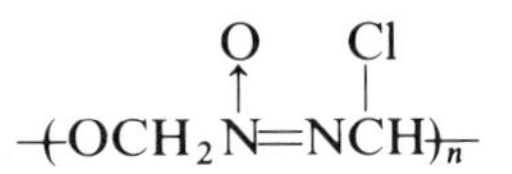 poly[oxymethylene-*ONN*-azoxy(chloromethylene)]

The unsymmetrical group —N=N—NH—, designated 'diazoamino' under the IUPAC organic nomenclature rules [3], in the present directional nomenclature for polymers is called 'azoimino'.

Among hetero atoms of the same kind, the shortest path and direction between the hetero atoms is chosen first. For a choice between equal paths, the hetero atom of highest seniority is the one most highly substituted, with the order of substituent seniority being that given in Rule 2.4.2.

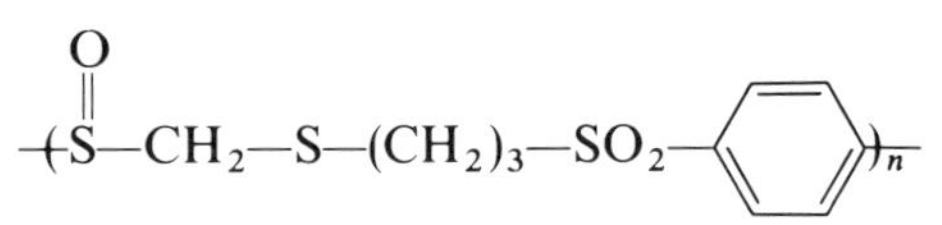 poly(sulfinylmethylenethiotrimethylenesulfonyl-1,4-phenylene)

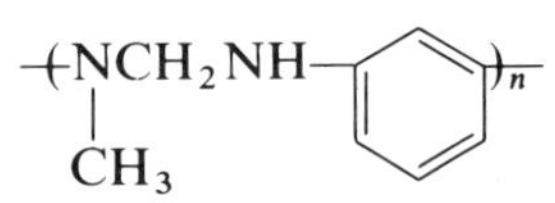 poly[(methylimino)methyleneimino-1,3-phenylene]

2.3.2

If the CRU contains two or more hetero atoms of highest seniority or more than two such hetero atoms separated by identical paths, the direction of citation is determined by the shorter path to the subunit of second seniority. Further choice is based on the shorter path from that subunit to the subunit of third seniority, etc as indicated in the order of seniority in Rule 2.3.1.

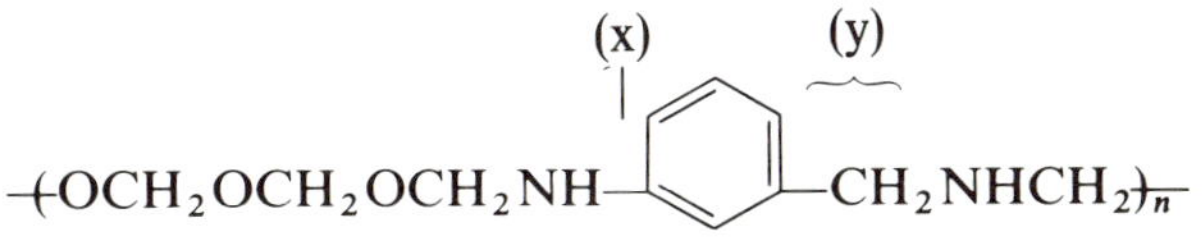

poly(oxymethyleneoxymethyleneoxymethyleneimino-1,3-phenylenemethyleneiminomethylene) or poly(1,3,5-trioxa-7-aza-heptamethylene-1,3-phenylene-2-azatrimethylene)
(the shorter path x between the NH group and the ring has been taken)

2.4 Carbocyclic rings and carbon chains

2.4.1

Constitutional repeating units in which the senior subunit is a carbocyclic ring system are named by citing first the carbocyclic ring of highest seniority and proceeding by the shorter path, in descending order of seniority, to (i) another of the same carbocycle; (ii) the carbocyclic system next in seniority; and (iii) the acyclic group appearing earliest in the alphabet. Carbocyclic ring system seniority is based on complexity, with the ring system of highest seniority being that containing the largest number of rings. Further order of seniority is based on (i) the largest individual ring at the first point of difference; (ii) the largest number of atoms common to the rings; (iii) the lowest locant numbers at the first point of difference for ring junctions; and (iv) the least hydrogenated ring. The basis for further choice is found in Rule C-14.1 of the IUPAC organic nomenclature rules [3]. The direction of citation in CRUs having two or more carbocycles of highest seniority is determined in a manner analogous to that of Rule 2.3.2.

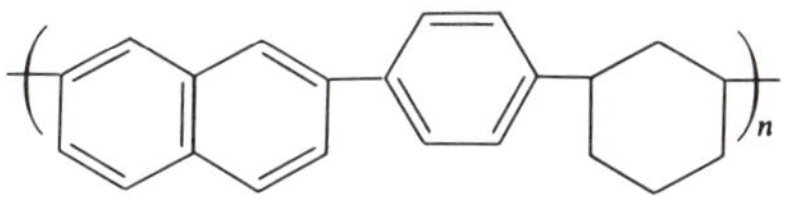

poly(2,7-naphthylene-1,4-phenylene-1,3-cyclohexylene)

Examples of the application of seniority rules in carbocyclic ring systems:

(i) Largest number of rings

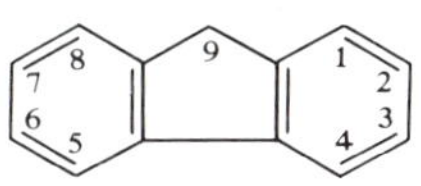

fluorene

is senior to

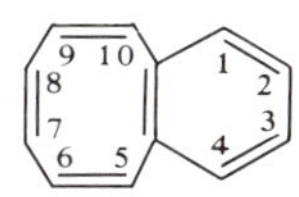

benzocyclooctene

(ii) Largest individual ring at the first point of difference

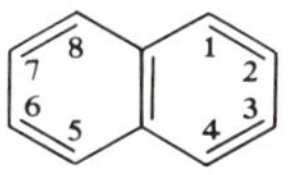

naphthalene

is senior to

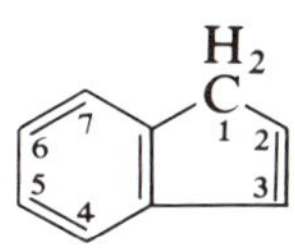

indene

(iii) Largest number of atoms common to the rings

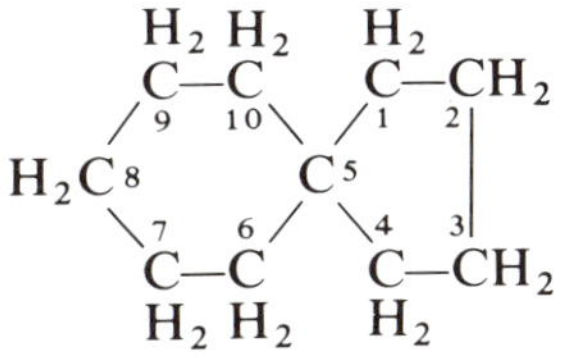

spiro[4.5]decane

is senior to

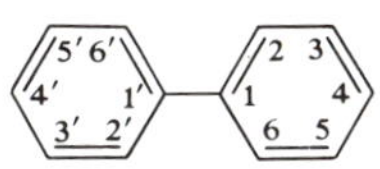

biphenyl

(iv) Lowest locant numbers at the first point of difference for ring junctions

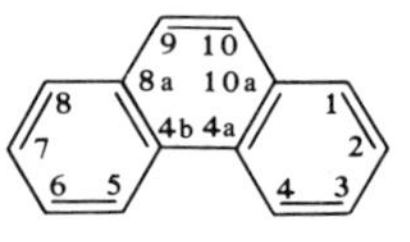

is senior to

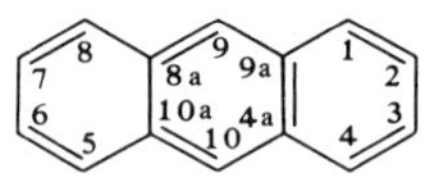

phenanthrene

anthracene

(v) Lowest state of hydrogenation

is senior to

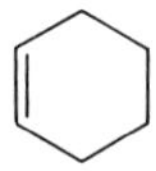

is senior to

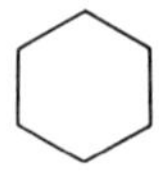

benzene

cyclohexene

cyclohexane

More than one numbering method may be in use in certain ring systems, such as the spiro hydrocarbons. Generally, in a specific ring system, a ring with unprimed locants is senior to one with primed locants. Points of attachment to the main chain of the CRU receive lowest permissible numbers.

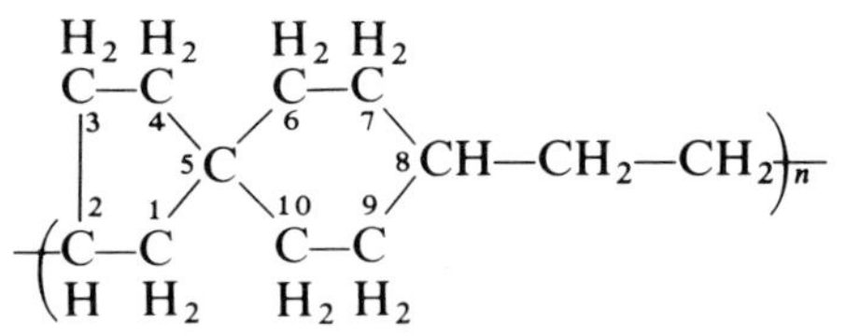

poly(spiro[4.5]dec-2,8-yleneethylene)

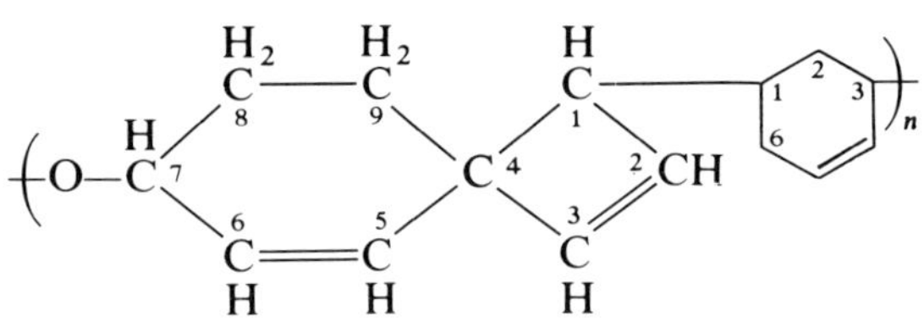

poly(oxyspiro[3,5]nona-2,5-dien-7,1-ylene-4-cyclohexen-1,3-ylene)

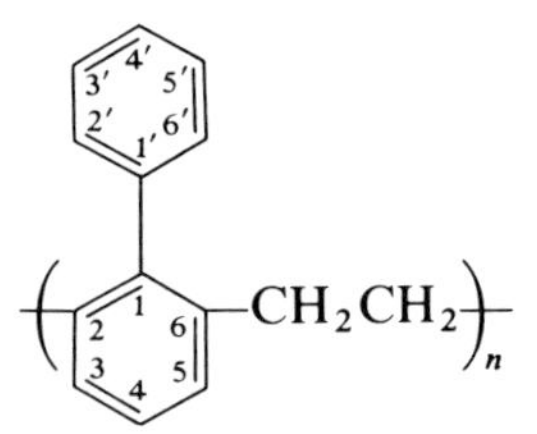

poly(2,6-biphenylyleneethylene)

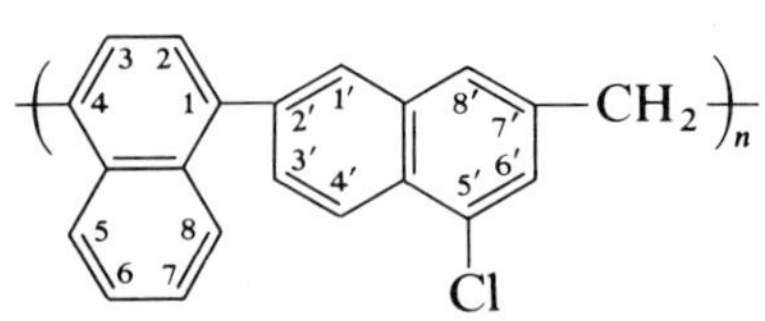

poly[(5′-chloro-1,2′-binaphthyl-4,7′-ylene)methylene]

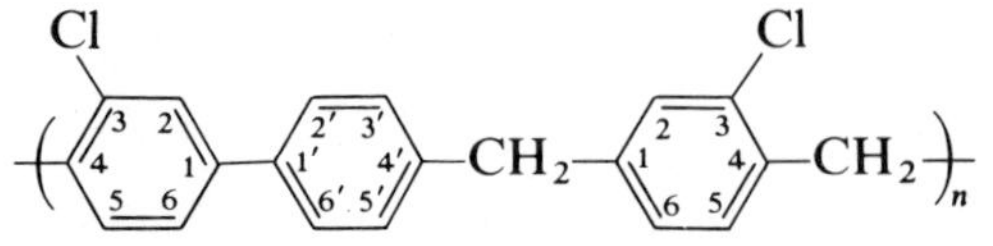

poly[(3-chloro-4,4′-biphenylylene)methylene(3-chloro-1,4-phenylene)methylene]

(not poly[(3′-chloro-4,4′-biphenylylene)methylene(2-chloro-1,4-phenylene)methylene]; the substituent in the preferred ring determines the direction)

2.4.2

When equal paths lead through two of the same acyclic subunits, choice of direction is determined, in descending order, by (i) the acyclic chain with the largest number of substituents

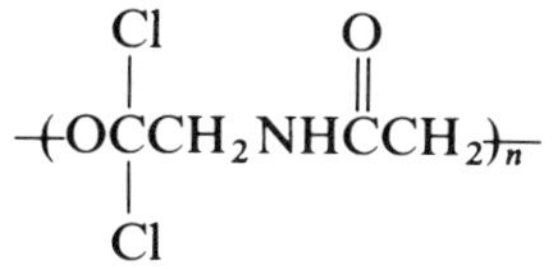

poly[oxy(1,1-dichloroethylene)imino(1-oxoethylene)]

(ii) the chain having substituents with lowest locants

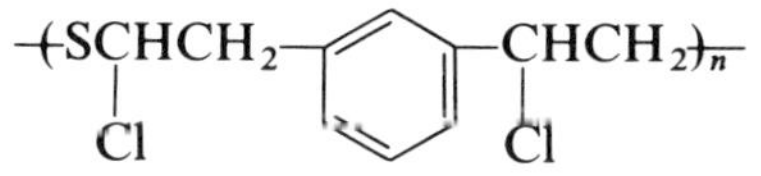

poly[thio(1-chloroethylene)-1,3-phenylene(1-chloroethylene)]

and (iii) the alphabetical order of substituents

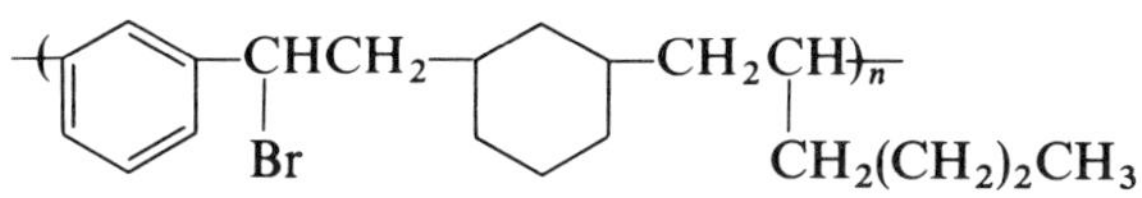

poly[1,3-phenylene(1-bromoethylene)-1,3-cyclohexylene(2-butylethylene)]

RULE 3 SUBSTITUENTS

3.1 Substituents included in trivial names

Substituents to acyclic or cyclic subunits in the main chain of the CRU are included within the trivial name of the subunit wherever such name is approved by the IUPAC organic rules [3] (see also Rule 1.2.2).

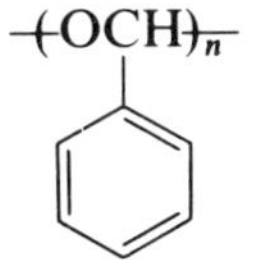

poly(oxybenzylidene)

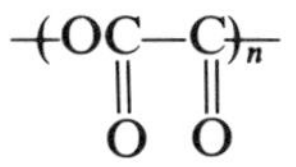

poly(oxyoxalyl)

3.2 Substituents named by prefixes

Substituents along the main chain other than those included in the name of a subunit are denoted by means of prefixes appended to the name of the subunit to which they are bound. In groups not having numbering fixed by other criteria, lowest locants appear at the left side of the group as written in the CRU (See also Rule 2.1.4).

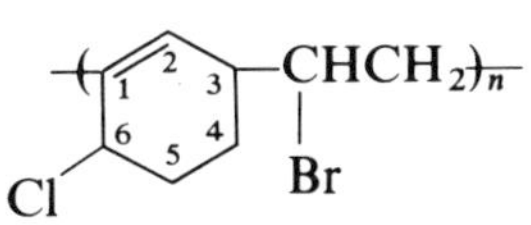

poly[(6-chloro-1-cyclohexen-1,3-ylene)(1-bromoethylene)]
(not poly[(6-chloro-1-cyclohexen-3,1-ylene)(2-bromoethylene)])

Functional derivatives that are clearly a part of the CRU are named as substituents to the appropriate subunit by the use of prefixes.

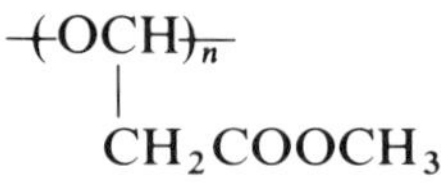

poly[oxy[2-(methoxycarbonyl)ethylidene]]

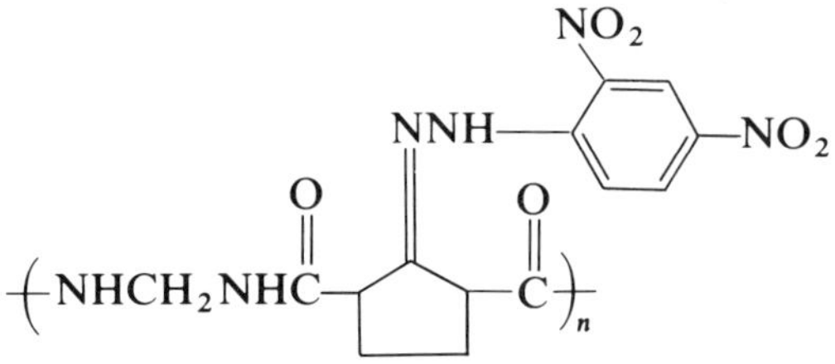

poly[iminomethyleneiminocarbonyl[2-[(2,4-dinitrophenyl)hydrazono]-1,3-cyclopentylene]carbonyl]

3.3 Salts and onium compounds of polymers

These are named by placing the appropriate prefix or suffix together with the name of the CRU in the enclosed part of the polymer name.

poly(sodium 1-carboxylatoethylene)

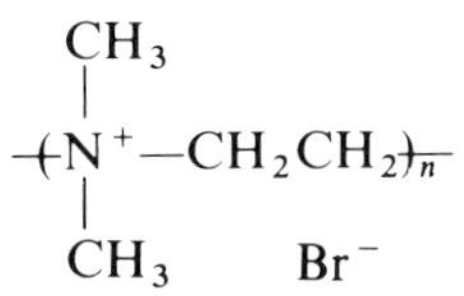

poly[(dimethyliminio)ethylene bromide]

Certain substituents are frequently expressed as part of a trivial name. The subunit thus named can itself be further substituted without altering the original trivial name.

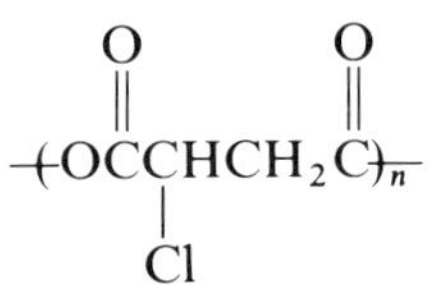

poly[oxy(2-chlorosuccinyl)]

The same (in this case, doubly-bonded oxygen) substituents not expressed in a trivial name have no special seniority.

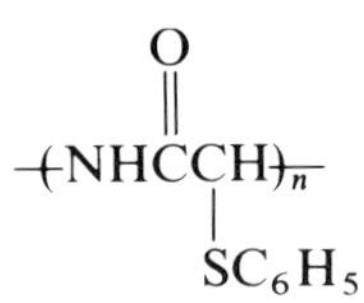

poly[imino[1-oxo-2-(phenylthio)ethylene]]

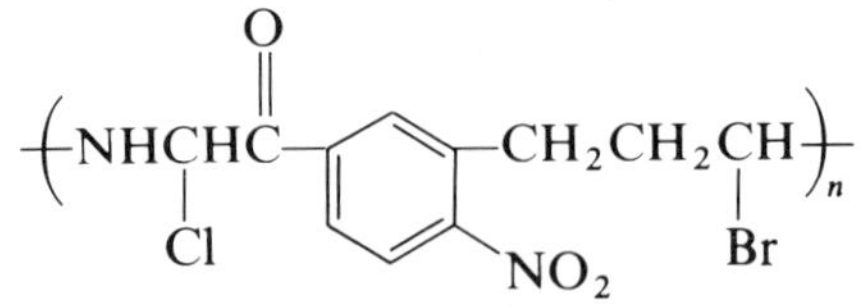

poly[imino(1-chloro-2-oxoethylene)(4-nitro-1,3-phenylene)(3-bromotrimethylene)]

Substituents in unknown positions in specific subunits are named in the usual way but either without locants or with an x locant.

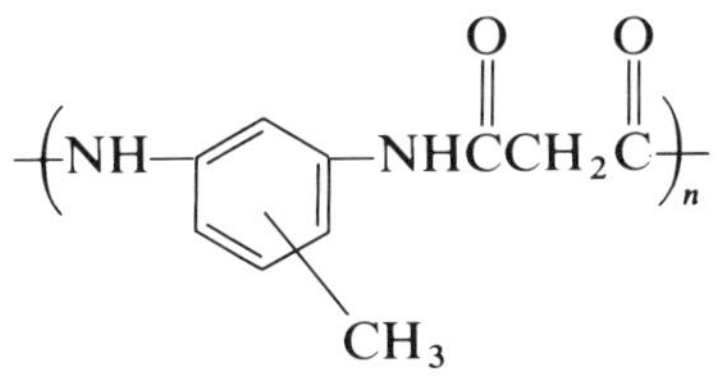

poly[imino(methyl(or x-methyl)-1,3-phenylene)iminomalonyl]

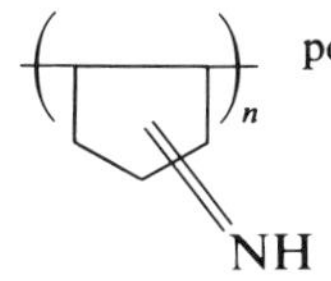

poly(x-imino-1,2-cyclopentylene)
(the x is required to differentiate this structure from 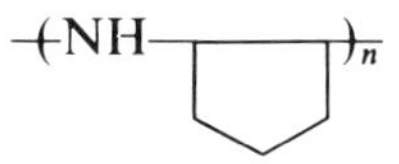

poly(imino-1,2-cyclopentylene))

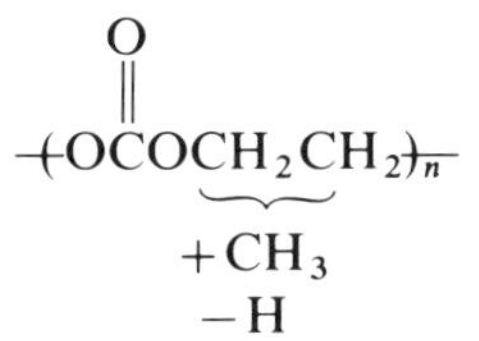

poly[oxycarbonyloxy(methylethylene)]
(position of the methyl not stated)

3.4 End groups

These may be specified by prefixes placed ahead of the name of the polymer. The end group designated by α is that attached to the left side of the CRU written as described in the preceding rules; the other end group is designated by ω.

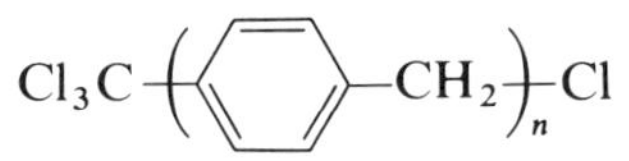

α-(trichloromethyl)-ω-chloropoly(1,4-phenylenemethylene)

Acknowledgement

The Commission acknowledges its debt to the Committee on Nomenclature of the Division of Polymer Chemistry of the American Chemical Society, whose efforts resulted in a general updating and extension of the 1952 IUPAC Polymer Rules [1]. The updated rules appeared in *Macromolecules* **1**, 193–198 (1968).

APPENDIX: SYSTEMATIC AND SOURCE NAMES FOR COMMON POLYMERS

The Commission recognized that a number of common polymers have semisystematic or trivial names that are well established by usage; it is not intended that they be immediately supplanted by the structure-based names. Nevertheless, it is hoped that for scientific communication the use of semisystematic or trivial names for polymers will be kept to a minimum.

For the following idealized structural representations, the semisystematic or trivial names given are approved for use in scientific work; the corresponding structure-based names are given as alternative names. Equivalent names for close analogues of these polymers (e.g. other alkyl ester analogues of poly(methyl acrylate)) are also acceptable. Where the semisystematic name is an obvious source-based name, the polymer referred to is that derived from the indicated source.

NOMENCLATURE

$-(CH_2CH_2)_n-$	polyethylene poly(methylene)
$-(CH(CH_3)CH_2)_n-$	polypropene poly(propylene)
$-(C(CH_3)_2CH_2)_n-$	polyisobutylene poly(1,1-dimethylethylene)
$-(CH{=}CHCH_2CH_2)_n-$	polyisobutylene poly(1,1-dimethylethylene)
$-(C(CH_3){=}CHCH_2CH_2)_n-$	polyisoprene poly(1-methyl-1-butenylene)
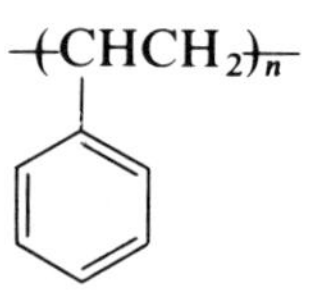	polystyrene poly(1-phenylethylene)
$-(CH(CN)CH_2)_n-$	polyacrylonitrile poly(1-cyanoethylene)
$-(CH(OH)CH_2)_n-$	poly(vinyl alcohol) poly(1-hydroxyethylene)
$-(CH(OCOCH_3)CH_2)_n-$	poly(vinyl acetate) poly(1-acetoxyethylene)
$-(CH(Cl)CH_2)_n-$	poly(vinyl chloride) poly(1-chloroethylene)
$-(CF_2CH_2)_n-$	poly(vinylidene fluoride) poly(1,1-difluoroethylene)
$-(CF_2CF_2)_n-$	poly(tetrafluoroethylene) poly(difluoromethylene)

Structure	Name
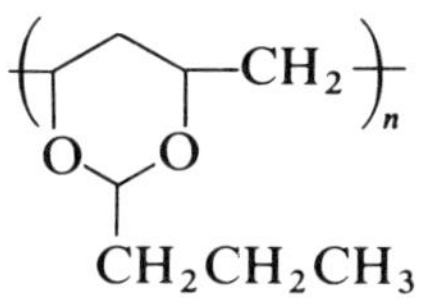	poly(vinyl butyral) poly[(2-propyl-1,3-dioxane-4,6-diyl)methylene]
$-(CH(COOCH_3)CH_2)_n-$	poly(methyl acrylate) poly[1-(methoxycarbonyl)ethylene]
$-(C(CH_3)(COOCH_3)CH_2)_n-$	poly(methyl methacrylate) poly[1-(methoxycarbonyl)-1-methylethylene]
$-(OCH_2)_n-$	polyformaldehyde poly(oxymethylene)
$-(OCH_2CH_2)_n-$	poly(ethylene oxide) poly(oxyethylene)
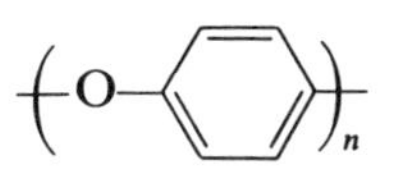	poly(phenylene oxide) poly(oxy-1,4-phenylene)
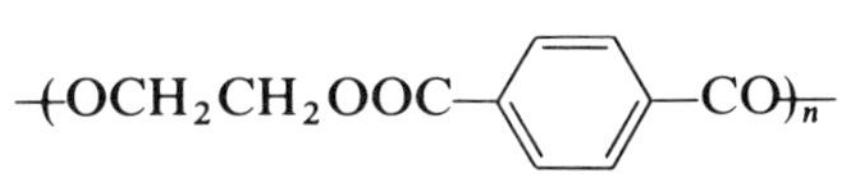	poly(ethylene terephthalate) poly(oxyethyleneoxyterephthaloyl)
$-(NHCO(CH_2)_4CONH(CH_2)_6)_n-$	poly(hexamethylene adipamide) poly[imino(1,6-dioxohexamethylene)iminohexamethylene] or poly(iminoadipolyiminohexamethylene)
$-(NHCO(CH_2)_5)_n-$	poly(ϵ-caprolactam) poly[imino(1-oxohexamethylene)]

REFERENCES

1 IUPAC. Report on nomenclature in the field of macromolecules. *J. Polymer Sci.* **8**, 257–277 (1952).
2 IUPAC. Report on nomenclature dealing with steric regularity in high polymers. *Pure Appl. Chem.* **12**, 643–656 (1966); *Bull. Soc. Chim. Fr.* **7**, 2127–2132 (1965).
3 IUPAC. *Nomenclature of Organic Chemistry*, Sections A, B, C, D, E, F and H. Pergamon Press, Oxford (1979).
4 IUPAC. Basic definitions of terms relating to polymers 1974. *Pure Appl. Chem.* **40**, 477–491 (1974). Reprinted as Chapter 1, this volume.

6: Nomenclature for Regular Single-Strand and Quasi-Single-Strand Inorganic and Coordination Polymers (1984)

PREAMBLE

The system for naming polymers in terms of structure previously published [1, 2] dealt with linear organic polymers, primarily those defined [2] as regular single-strand polymers, and followed as closely as possible established principles of organic nomenclature [3]. Accordingly, constituent subunits of the smallest repeating structural unit, named as bivalent radicals, are combined additively to form the name of the constitutional repeating unit. Extension of this method to linear inorganic and/or coordination polymers is seriously limited by the general lack of a system for naming bivalent radicals, because of the basic difference in philosophy between inorganic nomenclature and organic nomenclature. Furthermore, the constituent units of the constitutional repeating unit in most inorganic and coordination polymers are not bivalent radicals in the usual sense.

The present system is designed to name, uniquely and unambiguously, regular inorganic and/or coordination linear polymers, the constituent subunits of which can be formulated according to usual chemical principles of covalent and/or coordinate covalent bonding and the structures of which can be described by a constitutional repeating unit with at least one terminal constituent subunit that is connected through only one atom to other identical constitutional repeating units, or to an end group. 'Ladder' structures are thus excluded.

A *regular linear polymer* that can be described by a preferred constitutional repeating unit in which *both* terminal constituent subunits are connected through single atoms to the other identical constitutional repeating units or to an end group is called a *regular single-strand polymer* (see also Reference [2]).

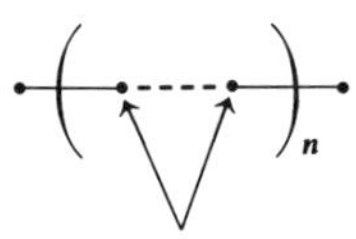

single atoms

A *regular linear polymer* that can be described by a preferred constitutional repeating unit in which *only one* terminal constituent subunit is connected through a single atom to the other identical constitutional repeating units or to an end group is a *quasi-single-strand polymer*, i.e. it does not fit the definition of regular single-strand polymers [2c], but can be named in the same manner.

Prepared by a working group consisting of L.G. Donaruma (USA), B.P. Block (USA), K.L. Loening (USA), N.A. Platé (USSR), T. Tsuruta (Japan), K. Ch. Buschbeck (FRG), W.H. Powell (USA) and J. Reedijk (Netherlands).
Reprinted from *Pure Appl. Chem.* **57**, 149–168 (1985).
Based on the document '*A Structure-Based Nomenclature for Polymers. II. One-Dimensional Inorganic and Semi-Organic Polymers*' prepared by the Nomenclature Committee of the American Chemical Society's Division of Polymer Chemistry (1977).

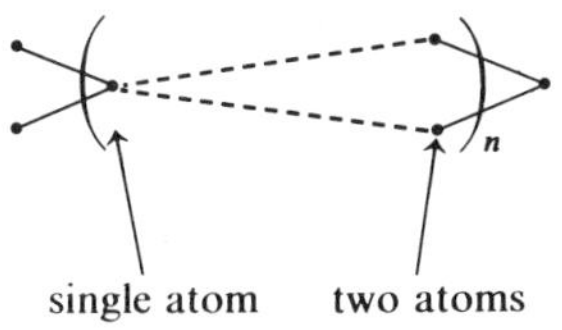

Established principles of inorganic and coordination nomenclature [4] are used as far as is consistent with the definitions [5] and basic principles [1, 2] of polymer nomenclature already published. As in the nomenclature of organic polymers, these rules apply to structures, not substances, that may be idealized representations of complex systems. Polymeric substances usually include a number of different structures, and a complete description of a polymer molecule would have to include such items as degree of steric regularity, chain imperfections, random branching, etc, resulting in extremely complex nomenclature. In any event, it is useful to consider a polymer in terms of a single structure that may itself be idealized. To the extent that an inorganic or coordination polymer can be represented as a linear combination of regularly repeating constitutional (structural) units, it can be named by the following rules. End groups may be included, if desired.

A few polymers with inorganic backbones in which the bonding is primarily covalent have trivial or semisystematic names of long standing, e.g. poly(dimethylsiloxane) for $\text{-(}Si(CH_3)_2\text{—}O\text{)}_n\text{-}$* and poly(dichlorophosphazene) for $\text{-(}PCl_2\text{=}N\text{)}_n\text{-}$†; some of these polymers can also be named by the principles for naming organic polymers, i.e. by using bivalent radicals [1, 2], e.g. poly[oxy-(dimethylsilylene)] for $\text{-(}Si(CH_3)_2\text{—}O\text{)}_n\text{-}$, and poly[nitrilo(dichlorophosphoranylidyne)] for $\text{-(}PCl_2\text{=}N\text{)}_n\text{-}$. There is no objection to the use of trivial or semisystematic names as long as they are clear and unambiguous, nor is there objection to the use of names based on the principles for naming organic polymers, if names for the bivalent radicals in the structure are clearly established. However, for some structures, the use of the rules given below provides unambiguous names with much less artificiality.

FUNDAMENTAL PRINCIPLES

The system of nomenclature for regular single-strand and quasi-single-strand inorganic and coordination polymers presented here is governed by the same fundamental principles of polymer nomenclature developed for single-strand organic polymers [2]. It is based on the selection and naming of a constitutional repeating unit (CRU), defined [5a] as the smallest structural unit the repetition of which describes the polymer structure. The name of the polymer is the name of this repeating unit prefixed by the terms 'poly', '*catena*', or other structural indicator, and designations for end groups, if desired.

The name of the CRU is formed by citing, in order of appearance along the chain of the CRU, the names of its constituent subunits, which are the *longest structural fragments* of the CRU that can be

*Trivial names of 'siloxane' polymers are based on a $\text{—}SiH_2\text{—}O\text{—}$ repeating unit named siloxane which, together with substituents, is enclosed in parentheses or brackets and preceded by the prefix 'poly'. Names of low-molecular-weight siloxane polymers can be named by using the prefix 'oligo' or a numerical prefix in place of 'poly' as described in the organic polymers rules [1, 2].

On the other hand, names of specific, low-molecular-weight, acyclic siloxanes with the general formula $H_3Si\text{-[}O\text{—}SiH_2\text{]}_n\text{-}O\text{—}SiH_3$ are formed according to Rule D-6.22 [3a]. This results in similar, yet not identical names; for example, when $n = 2$ in the general formula above, the name would be tetrasiloxane. In these names, substituents are cited as prefixes to such parent names, for example, decamethyltetrasiloxane.

†This name is a hybrid of additive and substitutive nomenclature that does not satisfy fully the rules of either system.

named by the established principles of inorganic and/or coordination nomenclature.* Accordingly, bridging ligands are not broken into subunits smaller than those named by principles of coordination nomenclature for ligands [4].

Although this procedure will result in an unambiguous name, it will not necessarily give a unique name. To obtain a unique name, a single, preferred CRU must be selected. This may be accomplished by the following procedure:

1. *identify* the CRU and its constituent subunits;
2. *orient* the CRU, i.e. determine the subunit that will begin the citation of the CRU and the direction to proceed along the backbone of the polymer chain in writing the rest of the CRU from left to right;
3. *name* the CRU in two basic steps: (i) *name the subunits* by established inorganic and/or coordination nomenclature principles; (ii) *assemble the names* of the subunits according to the preferred direction of citation to form the name of the CRU.

Note

It is important to identify and orient the CRU as far as possible before assembling the name of the CRU. However, occasionally the choice of a unique orientation will depend on names for individual subunits.

Identification of the constitutional repeating unit (CRU)

In many cases, the polymer structure is simple enough that the CRU and its constituent subunits can readily be identified.

Example

In the polymer

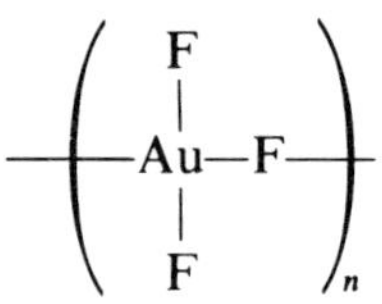

the two possible CRUs are:

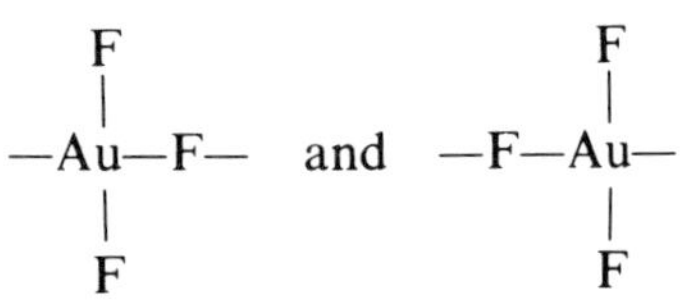

However, in more complex cases and occasionally in some simple cases it may be necessary to draw out a fairly long segment of the polymer chain in order to identify the possible constitutional repeating units.

Examples

1. In the polymer $\text{-(-Ag—CN—Ag—CN-)-}_n$, the possible CRUs are: —Ag—CN—; —Ag—NC—; —CN—Ag—; and —NC—Ag—. (As already stated the bridging ligand —CN— is not broken up into smaller subunits.)

*In the rules for naming organic polymers [1, 2], the constituent subunits of the CRU are the largest structural fragments that can be named as bivalent or multivalent radicals according to the established principles of organic nomenclature [3]. Single central atoms, mononuclear coordination centres, and bridging ligands are constituent subunits of CRUs in regular single-strand and quasi-single-strand inorganic and coordination polymers. Polynuclear coordination centres are used as subunits only under certain conditions (see discussion in 'Naming the constitutional repeating unit', p. 114).

2. In the polymer

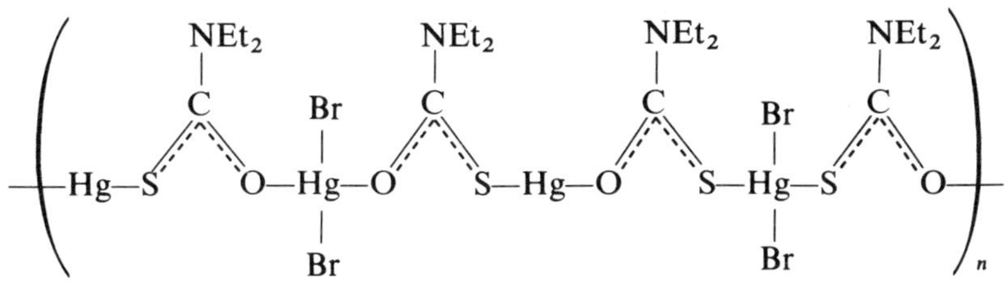

the possible constitutional repeating units are:

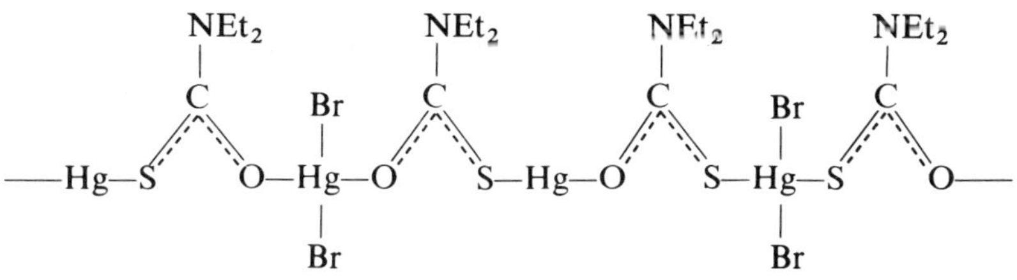

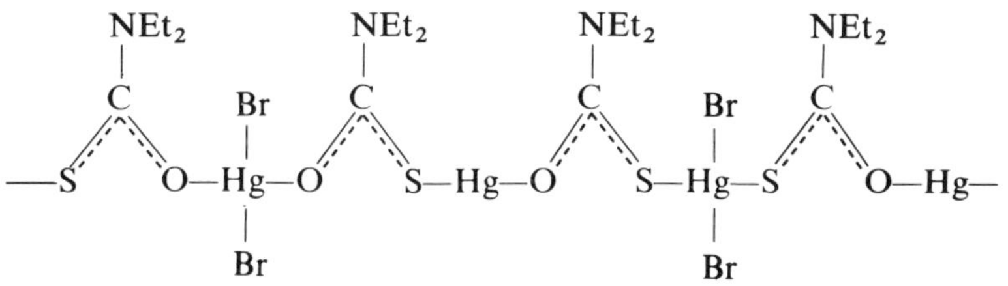

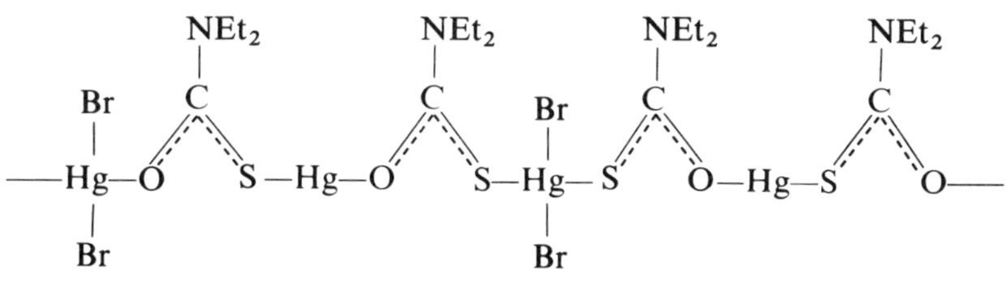

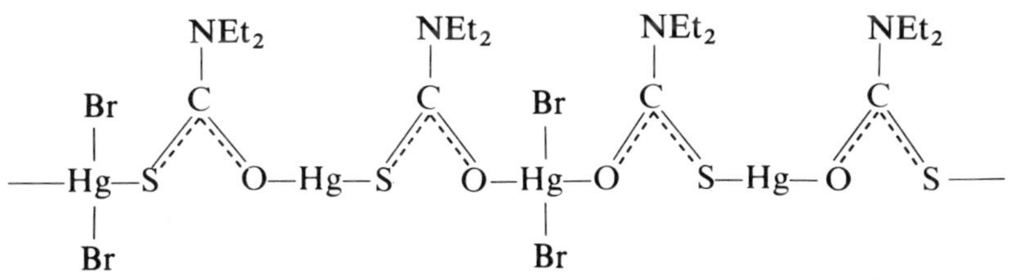

Orientation of the constitutional repeating unit

The constituent subunit of the CRU at which the citation of the CRU begins is the central atom (or coordination centre) of highest seniority, i.e. the most preferred central atom according to the set of hierarchical rules given in Rule IP-2.0, below. This centre is normally written as the *left* terminal subunit of the CRU.

The preferred direction along the polymer chain from the senior subunit for the sequential citation (from left to right) of the other constituent subunits in the CRU is governed by three major factors considered in order until a definitive decision is reached.

1. A single-strand CRU is preferred to a quasi-single-strand CRU, i.e. a CRU with *both* terminal constituent subunits connected to other identical constitutional repeating units or to an end group through single atoms is preferred to a CRU with only *one* terminal constituent subunit connected to other constitutional repeating units or to an end group through a single atom.
2. The preferred direction is defined by the shortest path, measured in terms of the number of atoms, in the polymer backbone from the senior subunit to a subunit of equal seniority, or to a subunit next in seniority.
3. When all paths between the senior subunit and a subunit of equal seniority, or a subunit ranking next in seniority, are of equal length, the preferred direction is along the path that includes constituent subunits of higher seniority. The paths between subunits of equal seniority or between the senior subunit and the subunit next in seniority necessarily involve subunits of lesser seniority, and often will consist of organic ligands. Hence, the hierarchical order of subunits prescribed for linear organic polymers [1, 2] may be needed to determine the preferred direction.

Further refinements to these general principles are given under Rule IP-2 and are illustrated in subsequent sections dealing with the naming of specific polymers.

Naming the constitutional repeating unit

The name of a CRU of a single-strand and quasi-single-strand inorganic or coordination polymers is based on a backbone consisting of central atoms and bridging ligands where present. All inorganic or coordination polymers have one or more central atoms, but may or may not have bridging ligands. Homoatomic inorganic polymers are considered to consist of central atoms only. Coordination centres, mononuclear or polynuclear, and their associated ligands, except for ligands between central atoms in the backbone, if any, are named by the usual principles of coordination nomenclature. Bridging ligands are named as ligands prefixed by the Greek letter μ.

Selection of the largest structural fragments in the backbone that can be assigned multivalent radical names as subunits of a CRU is a fundamental principle in naming linear organic polymers. For naming inorganic and coordination polymers, this principle is applied to the selection of bridging ligands in the CRU. When there is a choice, the largest group that can be named by the accepted methods for naming polydentate ligands is chosen. For example, in the polymer shown below, the CRU could be considered as two central atoms connected by

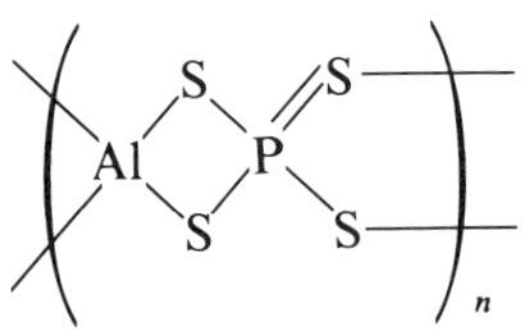

sulfur ligands. However, the principle of 'largest bridging ligand' requires the bridging ligand to be phosphorotetrathioato(3−). Strict application of this principle to inorganic or coordination polymers would lead to the selection of polynuclear coordination centres as the 'largest' structural fragment in the backbone. Since, at the present time, there are no officially accepted rules for uniquely naming and/or numbering certain types of polynuclear coordination centres, it is not yet convenient in some cases to use polynuclear coordination centres as subunits of a CRU in inorganic and coordination polymers. Hence, the principle of largest subunit is not always applied to coordination centres of a CRU and, in this set of rules, polynuclear coordination centres are used as subunits of CRUs only when it is *not* convenient to express such structural units in terms of their mononuclear coordination centres (see Rule IP-5). However, for illustrative purposes, names using polynuclear subunits are given as alternatives for some of the examples in the rules that follow.

Once the names of the constituent subunits of the CRU are determined, the CRU name is formed by citing the name of the senior subunit followed by the names of the other constituent subunits as they occur in the preferred direction along the polymer chain.

RULE IP-1 THE GENERAL POLYMER NAME

IP-1.1

A name of a polymer, for which the CRU is known, but with a dimensional structure* that may not be known, or that need not be specified, consists of the prefix 'poly' followed by the name of the constitutional repeating unit (see Rules IP-3–IP-7) enclosed in square brackets, e.g. poly[CRU].†

IP-1.2

If it is desired to specify the number of constitutional repeating units, the appropriate numerical prefix [3b] may be used in place of the prefix 'poly', e.g. deca[CRU].

IP-1.3

A linear (one dimensional) polymer is indicated by the italicized prefix '*catena*' added to the name of the polymer formed according to Rules IP-1.1 and IP-1.2‡, e.g. *catena*-poly[CRU].

IP-1.4

End groups of a polymer molecule may be specified, if desired, by appropriate prefixes identified by the Greek letters α and ω, which are added to the name of the polymer formed as given by Rules IP-1.1 or IP-1.2 and IP-1.3, e.g. α-(end group)-ω-(end group)-*catena*-poly[CRU]. For details see Rule IP-8.

RULE IP-2 SENIORITY RULES FOR SELECTION OF A PREFERRED CONSTITUTIONAL REPEATING UNIT

Many regular single-strand and quasi-single-strand inorganic or coordination polymers can be represented as multiples of a repeating unit most conveniently represented as a series of smaller subunits. The following rules are concerned with various seniority considerations necessary for the derivation of a preferred CRU. Refinements of these basic rules and their application to specific polymers are illustrated in the sections that follow.

IP-2.1 Choice of the senior constituent subunit

IP-2.1.1

The constituent subunit of highest seniority, i.e. the first subunit to be cited, in the preferred CRU of

*Linear (chain), crosslinked, branched, etc.

†Although the rules for naming linear organic polymers [1, 2] do not provide for describing dimensional structure, such specification would allow the prefix 'poly' to be a very general descriptor having no structural implications other than the presence of a number of constitutional repeating units.

‡The prefix '*catena*' is consistent with the existing rules for naming coordination compounds with 'extended' structures [4a, 4b]. In mineralogy and geochemistry, silicate chains have been denoted by the prefix '*ino*'; and the prefixes '*phyllo*' and '*tecto*' are used for sheet (two-dimensional) and three-dimensional structures, respectively [4c]. The term *catena* should not be confused with terms such as catenane, or catena compounds, used to describe interlocking organic ring compounds.

an inorganic or coordination polymer must contain one or more central atoms; bridging groups between central atoms in the backbone of the polymer cannot be senior subunits.*

IP-2.1.2

When there are two (or more) central atoms in a CRU of a linear inorganic or coordination polymer, the senior subunit is the one containing the central atom occurring latest in the general element sequence table (Fig. 6.1) [4d] shown below, as the line through the table is followed starting at the upper right corner.†

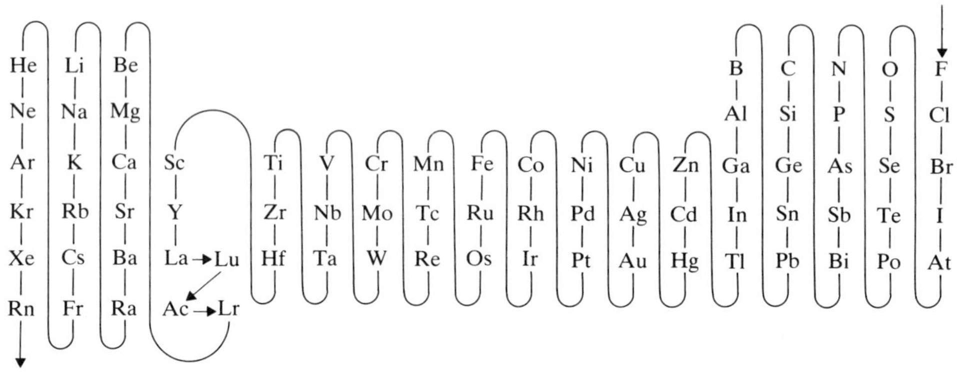

Fig. 6.1 Element sequence table.

IP-2.1.3

When a further choice is needed for the selection of a senior subunit in a CRU of a linear inorganic or coordination polymer, preference is given, in order, to:

1. a polynuclear coordination centre, in order of decreasing number of central atoms, provided the use of polynuclear centres as a subunit of a CRU is necessary (see discussion in 'Naming the constitutional repeating unit', p. 114);
2. the central atom or coordination centre with the greatest number of attached coordinating atoms, excluding coordinating atoms of bridging ligands in the backbone of the polymer chain;
3. the central atom or coordination centre the name of which, including ligands and their multiplying prefixes, if any, other than the bridging ligands in the backbone of the polymer chain, occurs earliest in the alphabet.

IP-2.2 Choice of the preferred direction along the polymer chain for the sequential citation of the constituent subunits of the CRU

After the senior subunits of the CRU have been determined by satisfying all of the seniority considerations of Rule IP-2.1, the following general principles are applied in order, where applicable.

IP-2.2.1

When it is possible to have either a single-strand or a quasi-single-strand CRU, the preferred

*This is consistent with the principles of coordination nomenclature in which the emphasis is always on the coordination centre. There is always at least one coordination centre in each inorganic or coordination polymer.

†Note that this seniority order is not the same as the seniority order for hetero atoms given in the organic polymer rules [2].

direction is that given a single-strand CRU*,

Example (M = central atom)

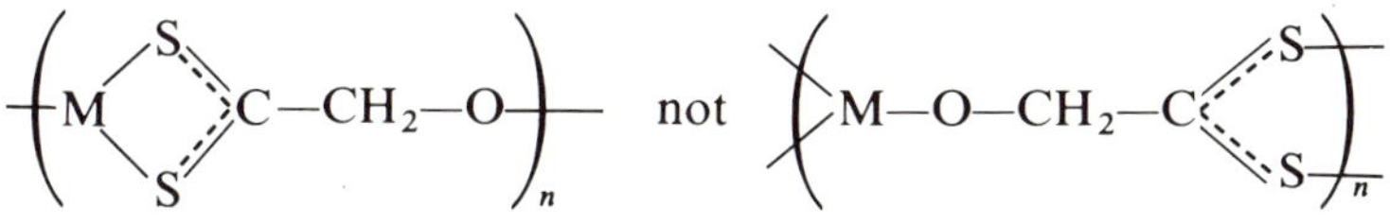

IP-2.2.2 Shortest path

The preferred direction along a polymer chain for the sequential citation of the constituent subunits of the CRU is the direction that leads first through the shortest path from the subunit of highest seniority to a subunit of equal seniority or to a subunit of next highest seniority. The length of the path between these subunits is the number of atoms in the most direct continuous chain of atoms from one unit to the other.

Examples (M = central atom)

1.

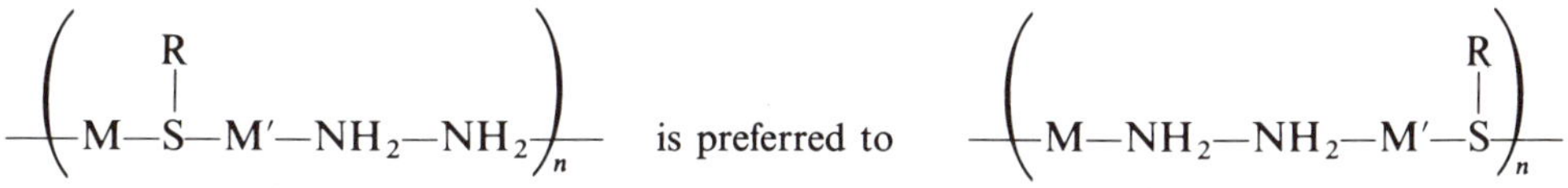

The one-atom path through the thiolato ligand is preferred to the two-atom path through the hydrazine ligand.

2.

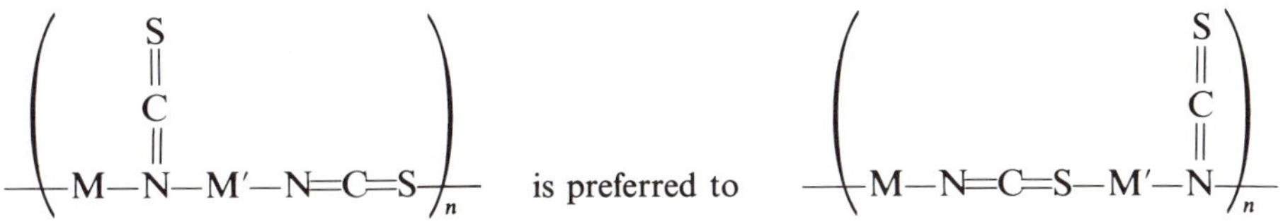

The one-atom path through the nitrogen atom of the isothiocyanato ligand is preferred to the three-atom path through all of the atoms of the isothiocyanato ligand.

3.

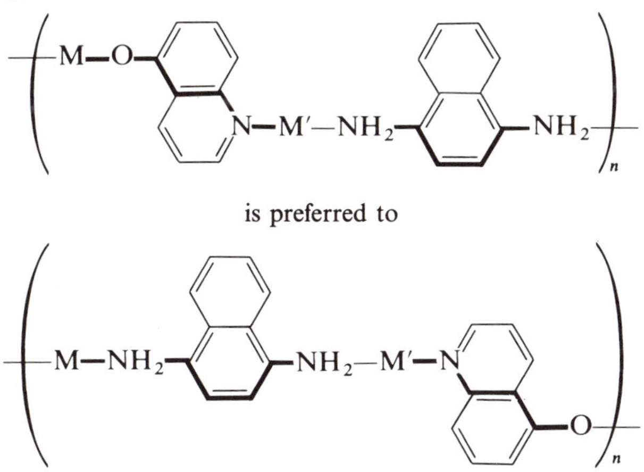

*This principle is quite analogous to that of minimizing free valences of constitutional repeating units in naming linear organic polymers when it is necessary to choose between a bivalent and a higher valent CRU after all factors concerned with the determination of subunit of highest seniority have been observed [2c].

The five-atom path through the 5-quinolinolato ligand is preferred to the six-atom path through the 1,4-naphthalenediamine ligand.

IP-2.2.3

When there are paths of equal (shortest) length between two subunits of equal highest seniority or between a subunit of highest seniority and a subunit ranking next in seniority, the preferred direction is determined by the kinds of structures and atoms included in the path and does not depend on actual names for subunits used in the final CRU name, unless there is no other choice remaining.

IP-2.2.3.1. For the selection of a preferred path, the same principles are used as in the organic polymer rules [2c], in which the fundamental order of seniority is (i) heterocycles; then (ii) acyclic hetero atoms; then (iii) carbocycles; and then (iv) acyclic carbon atoms or chains (see Rule 2.1.2, Reference [2c]). Seniority priority within these classes as given by the organic rules are also followed: for heterocycles*, see Rule 2.2.3 [2c]; for acyclic hetero atoms†, see Rule 2.3.1 [2c]; for carbocycles, see Rule 2.4.1 [2c]; for acyclic chains, see Rule 2.4.2 [2c].

Examples (M = central atom)

1.

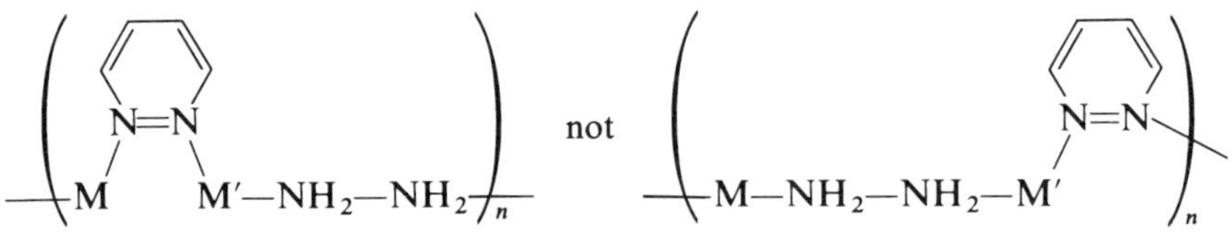

The heterocycle is preferred to the acyclic hetero chain.

2.

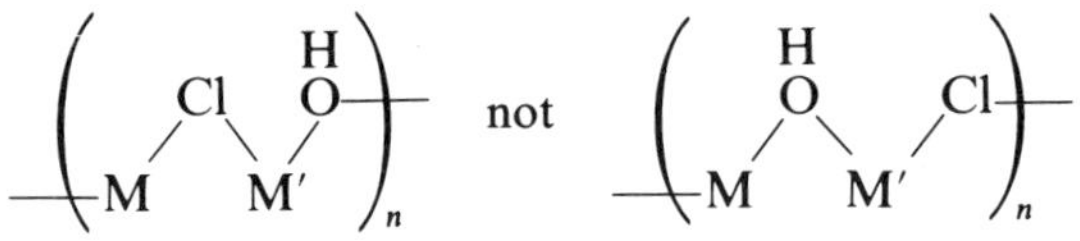

The Cl ligand is preferred to the O ligating atom.

IP-2.2.3.2. Substituents on atoms or groups in the path are used to determine priority between otherwise identical paths according to the principles in Rule 2.4.2 of the organic polymer rules [2c].

IP-2.2.3.3. If a further choice is necessary between otherwise identical paths, the preferred path leads by the shortest path, in the sense of Rule IP-2.2.2, from the senior subunit to the most preferred structural feature in the path.

Examples (M = central atom)

1.

$\left(\text{M—N}\equiv\text{C}\right)_n$ not $\left(\text{M—C}\equiv\text{N}\right)_n$

The hetero atom N is preferred to the carbon atom, and hence is preferred for citation closest to the senior subunit, M, in the preferred CRU.

*It is important to note that the seniority order for hetero atoms in ligands prescribed here is not the same as the seniority order for coordination centres, for which see Rule IP-2.1.2 in these rules.

2.

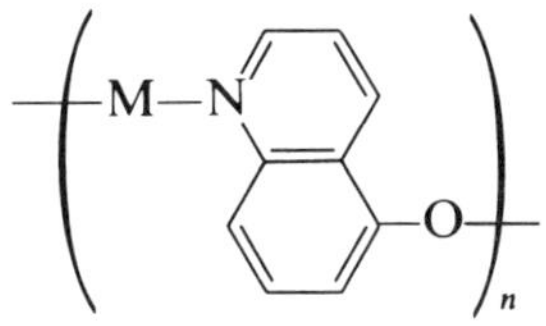

not

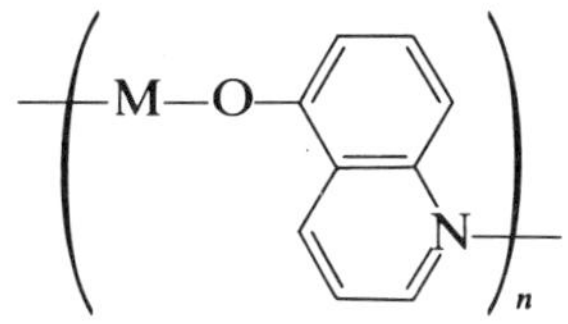

The heterocycle is preferred to the hetero atom O, and hence is preferred for citation closest to the senior subunit, M, in the preferred CRU.

IP-2.2.3.4. If a further choice is needed, the preferred path contains the ligand whose name occurs earliest in the alphabet.

RULE IP-3 REGULAR SINGLE-STRAND INORGANIC AND COORDINATION POLYMERS

Regular single-strand inorganic and coordination polymers are named by inserting the name of the preferred constitutional repeating unit into the appropriate general polymer name as given by Rule IP-1.

IP-3.1

Constitutional repeating units with homoatomic backbones are named by citing each mononuclear central atom, together with its side groups, if any, named as ligands.

*Examples**

1.
$\left(S \right)_n$ *catena*-poly[sulfur] [4e]

2.

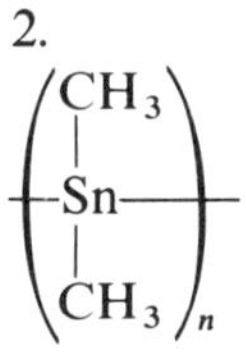

catena-poly[dimethyltin]

3.

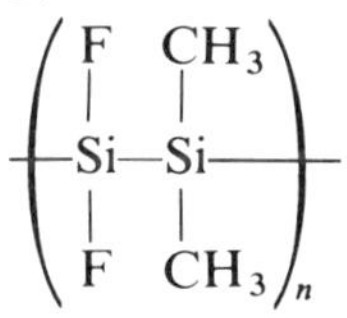

catena-poly[(difluorosilicon)(dimethylsilicon)]
(the subunit with the alphabetically earliest side group is the senior subunit)

IP-3.2

Constitutional repeating units with backbones consisting of a mononuclear central atom and one bridging ligand are named by citing the central atom prefixed by its associated non-bridging ligands followed by the name of the bridging ligand prefixed by the Greek letter μ.†

*According to the rules for naming linear organic polymers [1, 2], these inorganic homoatomic polymers could be named: (1) poly(sulfanediyl); (2) poly(dimethylstannylene) or poly(dimethylstannanediyl); (3) poly(1,1-difluoro-2,2-dimethyldisilane-1,2-diyl).

†Note that the bridging ligand is not included with the other ligands attached to the central atom. However, bridging ligands of polynuclear coordination centres that are not backbone units of the polymer are cited in their usual position in the name of the polynuclear coordination centre (see Rule IP-5).

Examples

1.

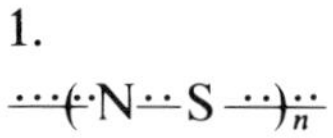

catena-poly[nitrogen-μ-thio]
(see Rule IP-3.2.1)

2.

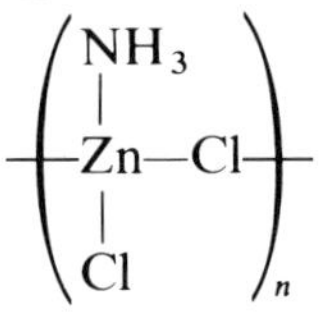

catena-poly[(amminechlorozinc)-μ-chloro]

3.

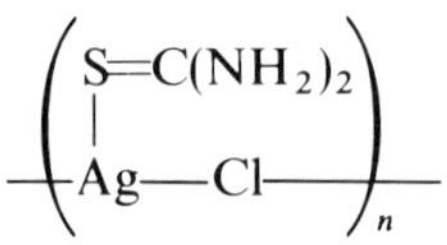

catena-poly[[(thiourea-*S*)silver]-μ-chloro]

IP-3.2.1

If there is a choice for the central atom, the element occurring later in the general element sequence table (see Rule IP-2.1.2) is the central atom.

Examples

1.

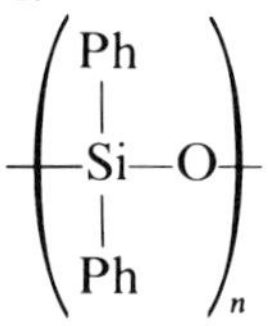

catena-poly[(diphenylsilicon)-μ-oxo]*

2.

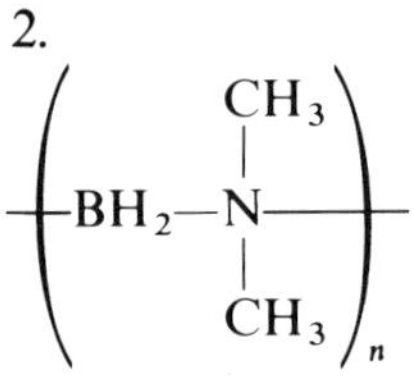

catena-poly[(dihydroboron)-μ-(dimethylamido)]

3.

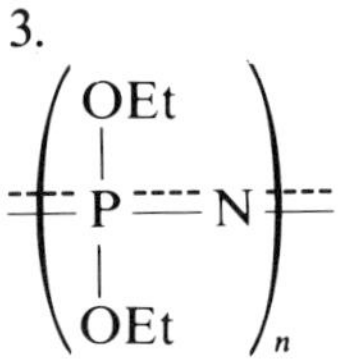

catena-poly[(diethoxophosphorus)-μ-nitrido]†

IP-3.2.2

Italicized element symbols indicating the coordinating atoms of bridging ligands in the backbone are cited in the order of the direction of citation of the CRU and are separated by a colon. Hence, element symbols cited before the colon refer to the central atom occurring just before the bridging

*According to the rules for linear organic polymers [1, 2], this polymer would be oriented and named poly[oxy(diphenylsilylene)].

†According to the rules for linear organic polymers [1, 2], this inorganic polymer would be oriented and named poly[nitrilo(diethoxyphosphoranylidyne)].

ligand in the CRU and element symbols cited after the colon refer to the central atom occurring immediately after the bridging ligand in the CRU or in the polymer chain.

Examples

1.
$\left(\!\!-Sn(CH_3)_3—S=C=N-\!\!\right)_n$ *catena*-poly[(trimethyltin)-μ-(thiocyanato-*S*:*N*)]

2.
$\left(\!\!-Ag—NC-\!\!\right)_n$ *catena*-poly[silver-μ-(cyano-*N*:*C*)]
(not *catena*-poly[silver-μ-(cyano-*C*:*N*)]; see Rule IP-2.2.3.3)

3.
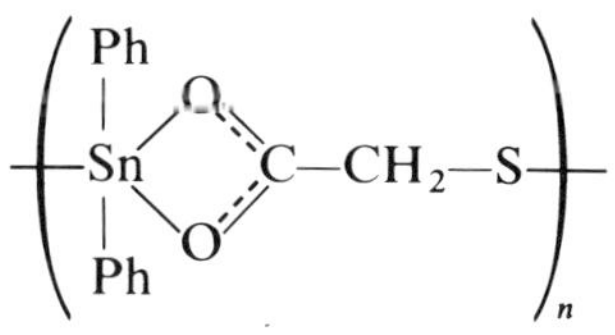

catena-poly[(diphenyltin)-μ-[mercaptoacetato(2−)-*O*,*O*′:*S*]
(direction of citation chosen by Rule IP-2.2)

When a choice of direction for citation of the constituent subunits in the CRU remains after application of the principles in Rule IP-2.2, the direction is chosen so that the italicized element symbols denoting coordinating atoms of the bridging ligand are cited in lowest alphabetic order (see Example 6 under Rule IP-3.3).

IP-3.3
Constitutional repeating units consisting of more than one central atom and no more than one bridging ligand between each central atom of the polymer, are named by extending the principles of Rule IP-3.2. The senior central atom is selected according to Rule IP-2.1 and the direction of citation is determined by Rule IP-2.2.

Examples

1.
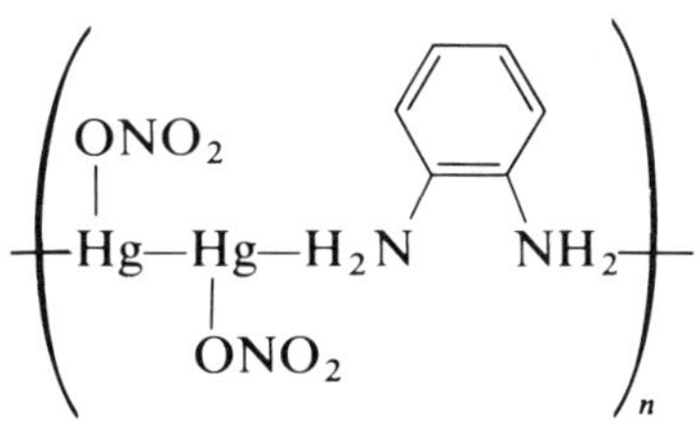

catena-poly[(nitratomercury)(nitratomercury)-μ-(*o*-phenylenediamine-*N*:*N*′)]
(not *catena*-poly[(nitratomercury)-μ-(*o*-phenylenediamine-*N*:*N*′)-(nitratomercury)]; Rule IP-2.2.2)*

2.
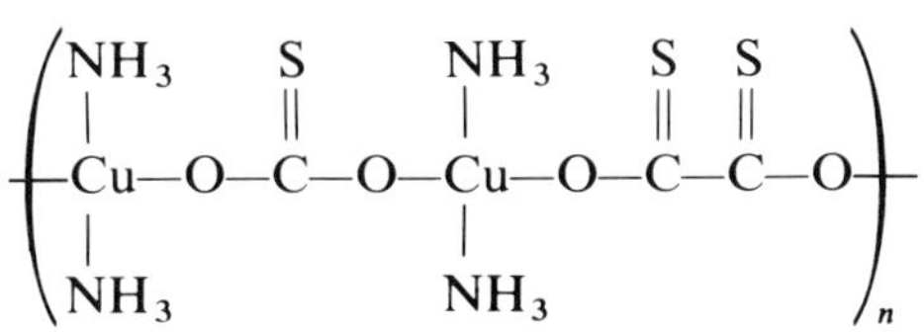

catena-poly[(diamminecopper)-μ-[thiocarbonato(2−)-*O*:*O*′]-(diamminecopper)-μ-[dithiooxalato(2−)-*O*:*O*′]]
(direction selected according to Rule IP-2.2.2)

*The omission of an indication for metal–metal bonding in this example is consistent with existing coordination rules for di- and polynuclear complexes [4f].

3.

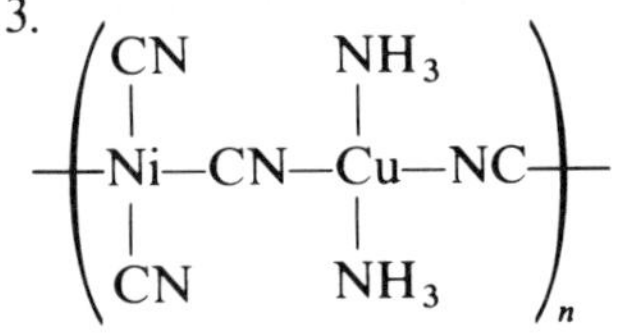

catena-poly[[bis(cyano-*C*)nickel]-μ-(cyano-*C*:*N*)-(diamminecopper)-μ-(cyano-*N*:*C*)]
(senior subunit chosen according to Rule IP-2.1.2)

4.

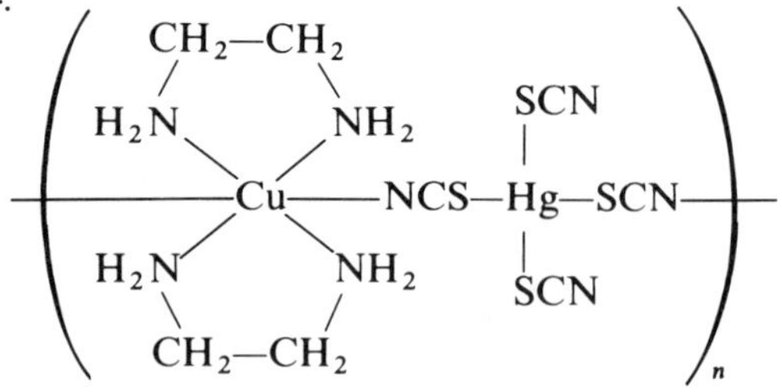

catena-poly[[bis(ethylenediamine-*N*,*N'*)copper]-μ-(thiocyanato-*N*:*S*)-[bis(thiocyanato-*S*)mercury]-μ-(thiocyanato-*S*:*N*)]

5.

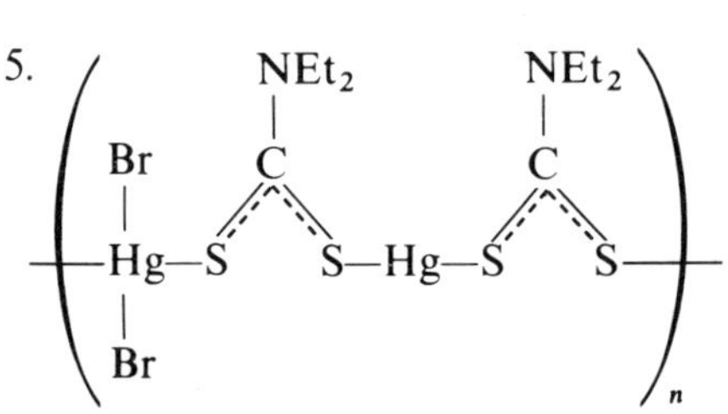

catena-poly[(dibromomercury)-μ-(diethyldithiocarbamato-*S*:*S'*)-mercury-μ-(diethyldithiocarbamato-*S*:*S'*)]
(senior subunit chosen according to Rule IP-2.1.3)

6.

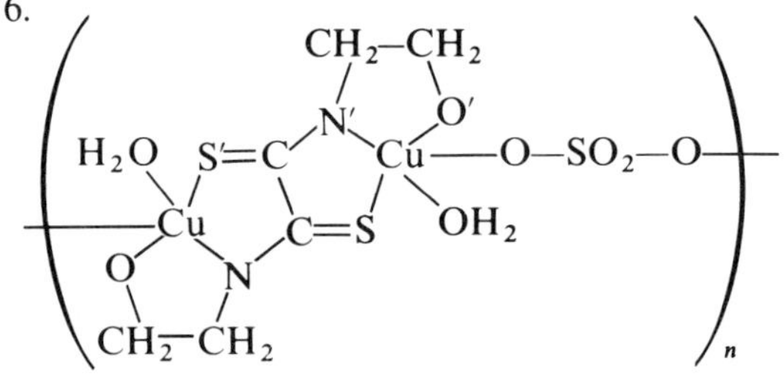

catena-poly[(aquacopper)-μ-[*N*,*N'*-bis(2-hydroxyethyl)dithiooxamido(2−)-*N*,*O*,*S'*:*N'*,*O'*,*S*]-[aquacopper)-μ-[sulfato-(2−)-*O*:*O'*]]*

*Orientations of the CRU that begin with either copper central atom and proceed first through the sulfato ligand are ruled out by Rule IP-2.2.1. These orientations produce the less preferred quasi-single-strand polymeric structure as shown below.

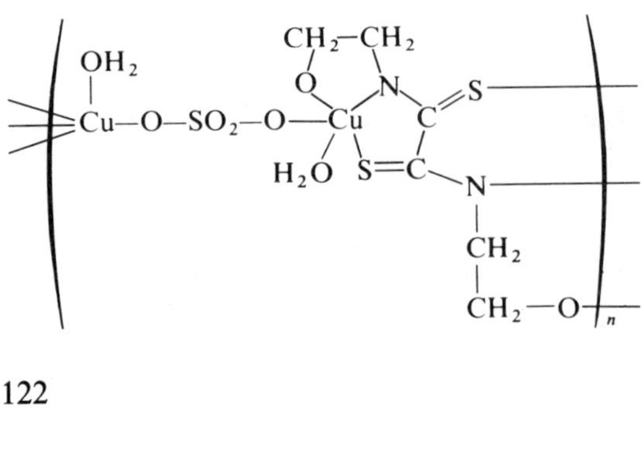

Multiple bridging ligands between the same pair of central atoms are cited in alphabetic order each preceded by the Greek letter μ, and all are enclosed in square brackets to reduce the possibility of misinterpretation.

7.

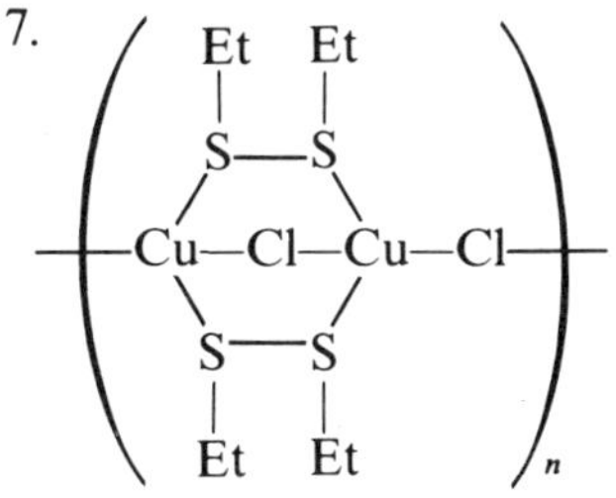

catena-poly[copper-[μ-chloro-bis-μ-(diethyl disulfide-*S*:*S*′)]-copper-μ-chloro]
(direction of citation chosen by Rule IP-2.2.1)*

8.

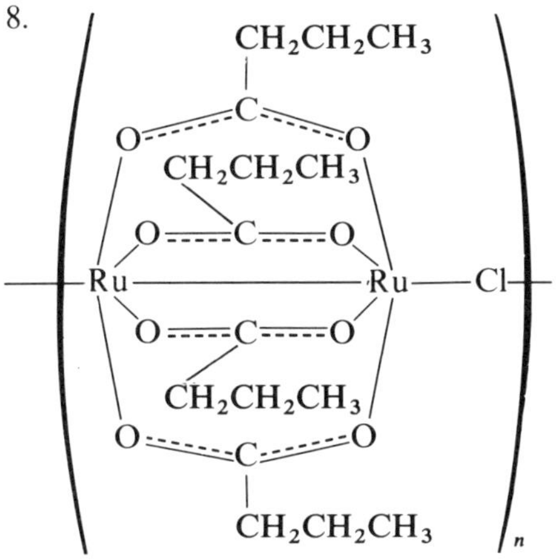

catena-poly[[ruthenium-tetrakis-μ-(butyrato-*O*:*O*′)-ruthenium(*Ru*-*Ru*)]-μ-chloro]
(direction of citation chosen by Rule IP-2.2.1)†

RULE IP-4 REGULAR QUASI-SINGLE-STRAND COORDINATION POLYMERS

Regular quasi-single-strand coordination polymers are named by inserting the name of the preferred constitutional repeating unit into the appropriate general polymer name as given by Rule IP-1.

Footnote *continued*

The choice between the remaining two orientations is made according to the second paragraph of Rule IP-3.2.2. In the preferred orientation, the letter locant order for the bridging ligand, *N*,*O*,*S*′:*N*′,*O*′,*S*, is lower than *N*′,*O*′,*S*:*N*,*O*,*S*′.

*It might be convenient in such cases to treat the two copper central atoms as a binuclear complex as follows: *catena*-poly[[μ-chloro-bis-μ-(diethyl disulfide-*S*:*S*′)-dicopper]-μ-chloro].

†It does not seem quite consistent with the philosophy of polymer nomenclature to cite a metal–metal bond in this fashion; hence, it would probably be more acceptable here to treat the two ruthenium central atoms as a binuclear complex, perhaps as follows: *catena*-poly[[tetrakis-μ-(butyrato-*O*:*O*′)-diruthenium(*Ru*-*Ru*)]-μ-chloro].

IP-4.1
Constitutional repeating units with backbones consisting of one mononuclear central atom and two or more bridging ligands, alike or different, or a chelating ligand, are named by citing the name of the central atom, prefixed by the names of its associated non-bridging ligand(s), followed by the names of the bridging ligands each prefixed by the Greek letter μ. The number of identical bridging ligands, if more than one, is indicated by an appropriate numerical prefix; different bridging ligands are cited in alphabetic order and all are enclosed in appropriate enclosing marks.

Examples

1.
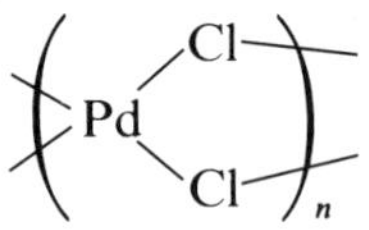

catena-poly[palladium-di-μ-chloro]
(*catena*-di-μ-chloro-palladium in the 1970 IUPAC Inorganic Rules [4g])

2.
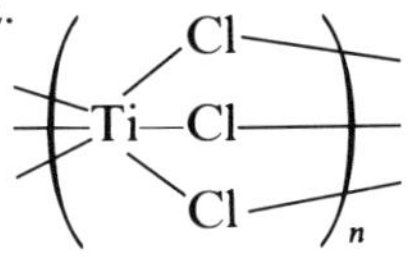
catena-poly[titanium-tri-μ-chloro]

3.
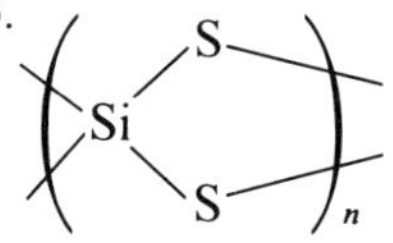
catena-poly[silicon-di-μ-thio]

4.
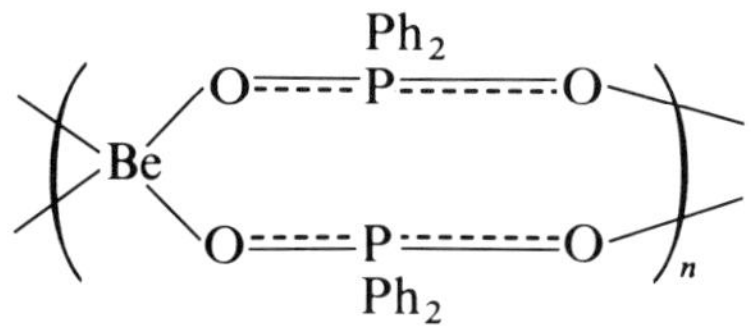
catena-poly[beryllium-bis-μ-[diphenylphosphinato(1 −)-*O*:*O*′]]

5.
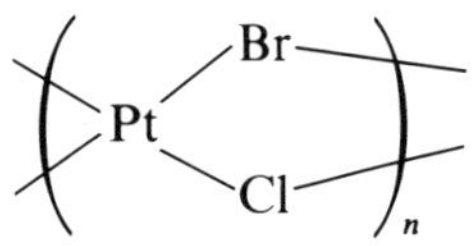
catena-poly[platinum(μ-bromo-μ-chloro)]

6.
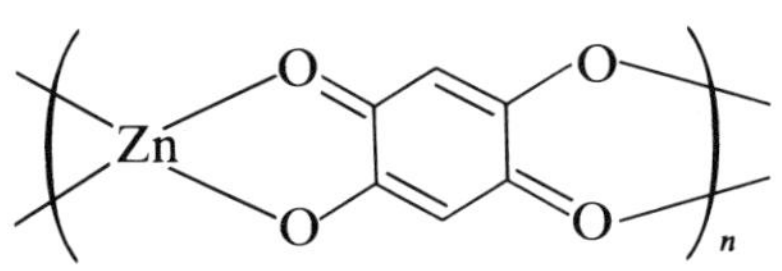

catena-poly[zinc-μ-[2,5-dihydroxy-*p*-benzoquinonato(2 −)-O^1,O^2:O^4,O^5]]

7.
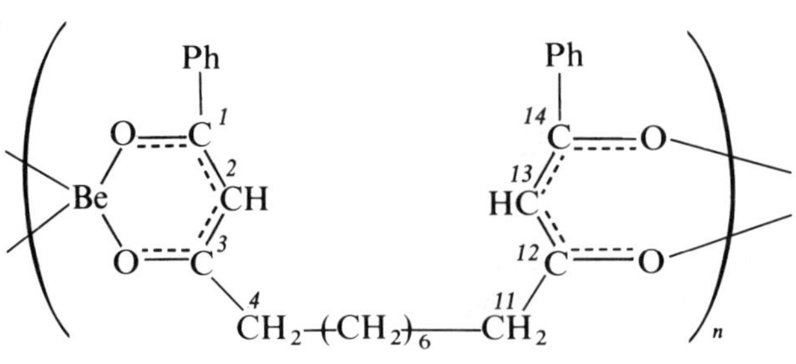

catena-poly[beryllium-μ-[1,14-diphenyl-1,3,12,14-tetradecanetetronato(2 −)-O^1,O^3:O^{12},O^{14}]]

8.
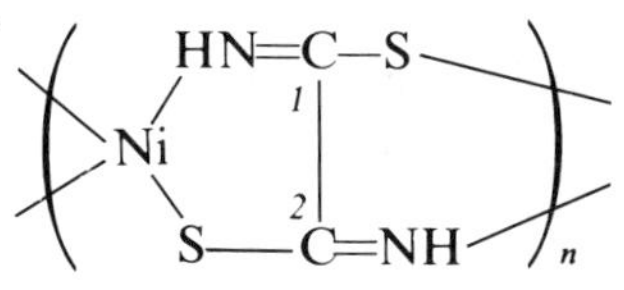

catena-poly[nickel-μ-[dithiooximidato(2−)-N^1,S^2:N^2,S^1]]

9.
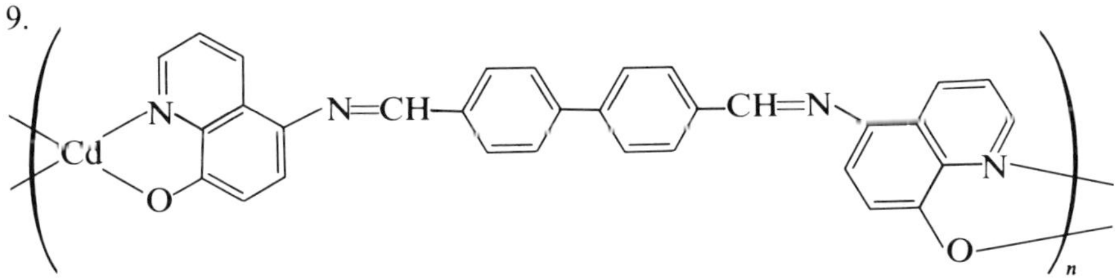

catena-poly[cadmium-μ-[[5,5′-[4,4′-biphenylylenebis(methylidynenitrilo)]di-8-quinolinolato](2−)-N^1,O:$N^{1'}$,O']]

IP-4.2

Constitutional repeating units containing more than one central atom are named by extending the principles of IP-4.1. The senior central atom is selected according to Rule IP-2.1, and the direction of citation is determined by Rule IP-2.2.

Examples

1.
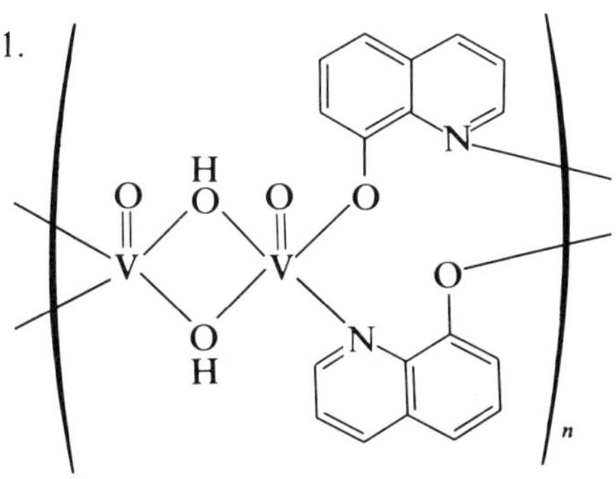

catena-poly[(oxovanadium)-di-μ-hydroxo-(oxovanadium)-[μ-(8-quinolinolato-N:O)-μ-(8-quinolinolato-O:N)]]
(the direction of citation is determined by Rule IP-2.2.2)

2.
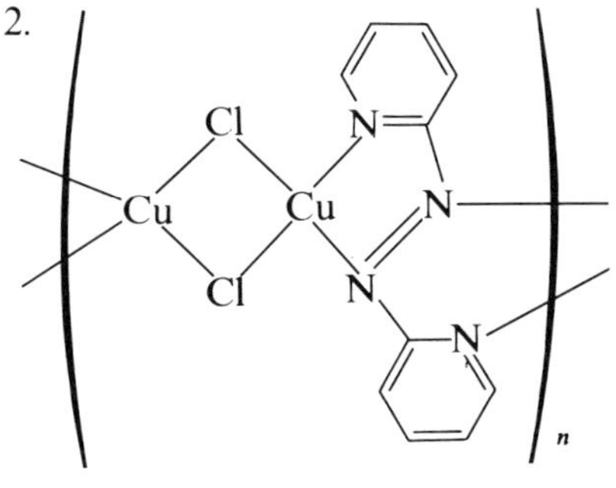

catena-poly[copper-di-μ-chloro-copper-μ-[2,2′-(azo-N:N')dipyridine-N':N]]
(the direction of citation is determined by Rule IP-2.2.2)

RULE IP-5 SINGLE-STRAND AND QUASI-SINGLE-STRAND COORDINATION POLYMERS WITH POLYNUCLEAR COORDINATION CENTRES

Single-strand and quasi-single-strand coordination polymers with one polynuclear coordination centre are named in much the same manner as coordination polymers having only mononuclear centres. The polynuclear centre is the senior subunit and begins the citation of the subunits in the name of the CRU. Both positions on the polynuclear complex where the bridging ligand is attached are indicated by numerical locants, inserted between the name of the polynuclear centre and the names of the bridging ligand and separated by a colon; locants before the colon refer to the preceding polynuclear centre in the CRU and locants following the colon refer to the next polynuclear centre in the polymer chain.

Examples

1. $\left[\left[\textit{octahedro}\text{-}W_6(\mu\text{-}Br_8)(2,3,4,5\text{-}Br_4)\right](6{:}1)(Br_4)\right]_n$

catena-poly[(octa-μ-bromo-2,3,4,5-tetrabromo-*octahedro*-hexatungsten)-6:1-μ-[tetrabromido(2−)]]

2. 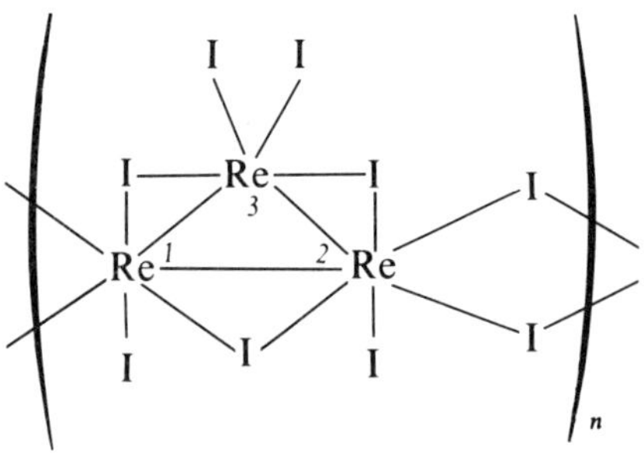

catena-poly[[1,2:1,3:2,3-tri-μ-iodo-1,2,3,3-tetraiodo-*triangulo*-trirhenium(3*Re-Re*)]-2,2:1,1-di-μ-iodo]*

RULE IP-6 REGULAR SINGLE-STRAND AND QUASI-SINGLE-STRAND INORGANIC AND COORDINATION POLYMERS WITH IONIC CONSTITUTIONAL REPEATING UNITS

Regular single-strand and quasi-single-strand inorganic and coordination polymers with ionic CRUs are named in the same general manner as described in Rules IP-3, IP-4 and IP-5. The charge of the CRU may be indicated by an Ewens–Bassett number cited after the name of the ionic portion of the CRU. Stock numbers may be used to denote the oxidation state of the central atom; if so, they are attached to the name of the central atom in the usual manner.

*Numbering rules for polynuclear complexes have not yet been fully defined and are under study by the IUPAC Commission on Nomenclature of Inorganic Chemistry. The numbering shown here is arbitrary and only for convenience in defining the structure of this polymer for this document.

*Examples**

1.

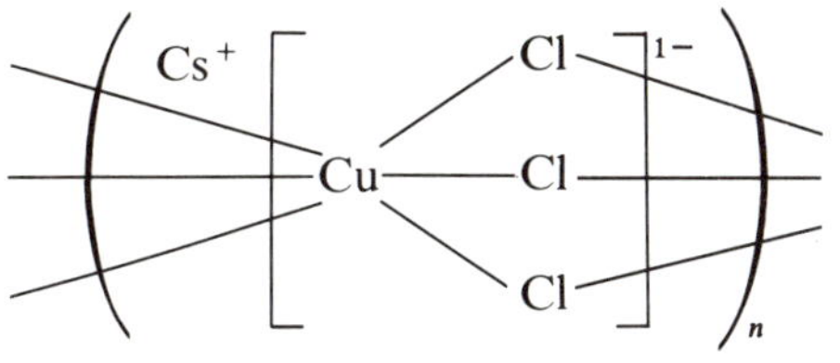

catena-poly[caesium [cuprate-tri-μ-chloro](1—)]

or

catena-poly[caesium [cuprate(II)-tri-μ-chloro]]

2.

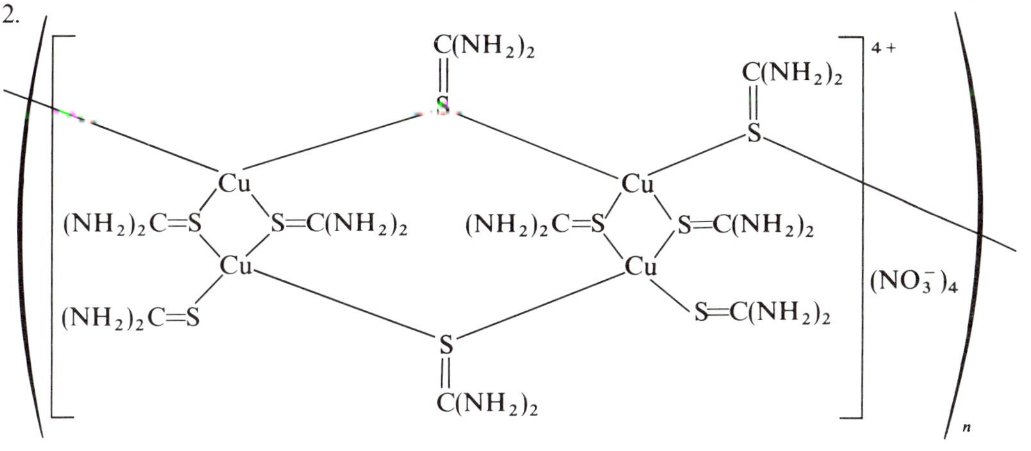

catena-poly[[[bis-μ-(thiourea-*S*,*S*)-2-(thiourea-*S*)dicopper]-1,2:1,2-[bis-μ-(thiourea-*S*:*S*)][bis-μ-(thiourea-*S*,*S*)-2-(thiourea-*S*)dicopper]-1:1-μ-(thiourea-*S*:*S*)](4+) tetranitrate]

or

catena-poly[[[bis-μ-(thiourea-*S*,*S*)-2-(thiourea-*S*)dicopper(I)]-1,2:1,2-[bis-μ-(thiourea-*S*:*S*)][bis-μ-(thiourea-*S*:*S*)-2-(thiourea-*S*)dicopper(I)-1:1-μ-(thiourea-*S*:*S*)] tetranitrate]

RULE IP-7 STEREOCHEMICAL CONFIGURATION OF A CONSTITUTIONAL REPEATING UNIT

Stereochemical configuration of a constitutional repeating unit consisting of a mononuclear central atom and one bridging ligand may be designated by suitable prefixes cited before the appropriate complete polymer name.†

*From an inorganic viewpoint, it might be better to consider these polymers to be salts of polymeric ions as illustrated below:

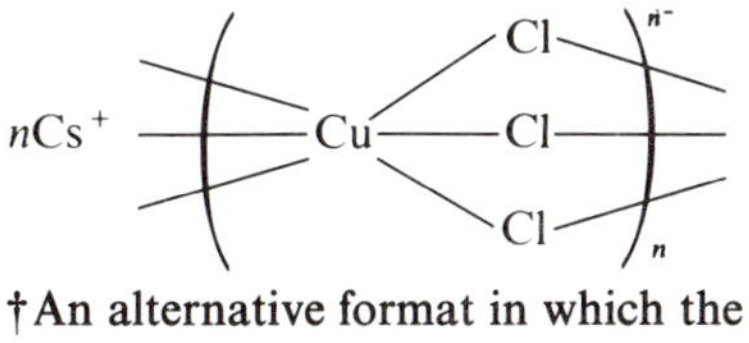

caesium *catena*-poly[[cuprate-tri-μ-chloro](1 −)]

or

caesium *catena*-poly[cuprate(II)-tri-μ-chloro]

†An alternative format in which the stereochemical prefix is inserted between the prefix 'poly' and the name of the CRU, as shown below for both examples under this rule, would be more consistent with inorganic nomenclature practice, as shown by the names below.

1. *catena*-poly[*cis*-[(difluorogold)-μ-fluoro]]
2. *catena*-poly[*trans*-[dipotassium [[(tetrafluoroaluminate)-μ-fluoro](2 −)]]]

   or

   catena-poly[*trans*-[dipotassium [[tetrafluoroaluminate(III)]-μ-fluoro]]]

However, the recommendation in this Rule (IP-7) is in accord with the stereochemical notation recommended for organic polymers [6].

Examples

1. 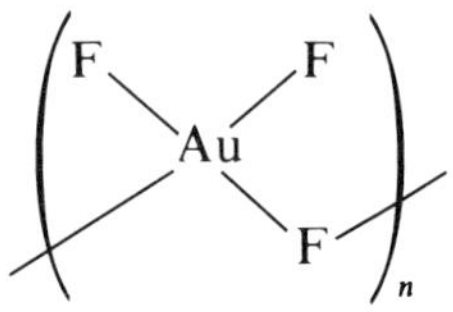

cis-catena-poly[(difluorogold)-μ-fluoro]

2. 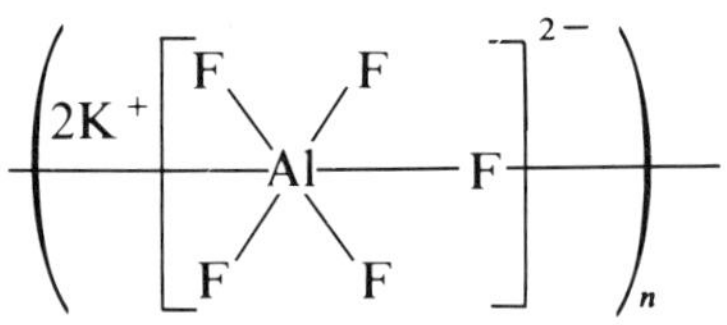

trans-catena-poly[dipotassium [[(tetrafluoroaluminate)-μ-fluoro](2−)]]
or
trans-catena-poly[dipotassium [[tetrafluoroaluminate(III)]-μ-fluoro]]
(inorganic nomenclature practice would not require use of the Stock number for aluminium)

RULE IP-8 END GROUPS OF LINEAR INORGANIC OR COORDINATION POLYMERS

End groups of linear inorganic or coordination polymers are specified by prefixes cited in front of the name of the polymer (see Rule IP-1.4).

IP-8.1

The groups attached to the first constituent subunit of the preferred CRU, i.e. the senior coordination centre written as the left terminal subunit in the CRU, are named as ligands and designated by the Greek letter α.

End groups attached to the other terminal subunit of the preferred CRU are named as ligands if attached to a central atom, or, if attached to a bridging ligand, are named as a central atom by the usual principles of coordination nomenclature and are designated by the Greek letter ω.

Examples

1. $Cl\text{-}(S)_n\text{-}H$ α-chloro-ω-hydro-*catena*-poly[sulfur]

2. 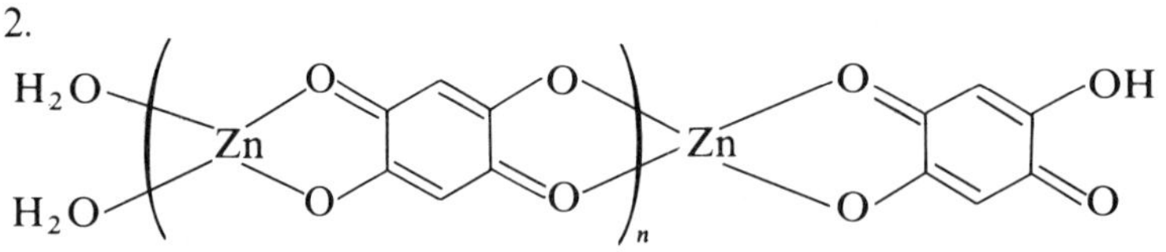

α,α-diaqua-ω-[[2,5-dihydroxy-*p*-benzoquinonato(1−)-*O*[1],*O*[2]]zinc]-*catena*-poly[zinc-μ-[2,5-dihydroxy-*p*-benzoquinonato-(2−)-*O*[1],*O*[2]:*O*[4],*O*[5]]]

IP-8.2

When a choice is necessary to determine ligands to be cited as an end group or to be included in the constitutional repeating unit, the ligand chosen as the α end group is the one whose name occurs earliest in the alphabet.

Example

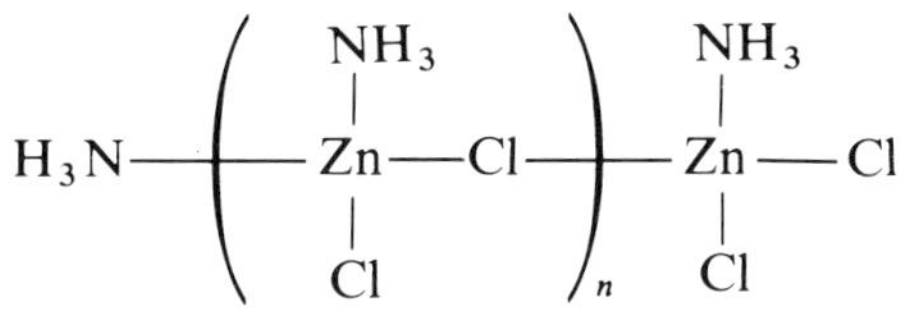

α-ammine-ω-(amminedichlorozinc)-*catena*-poly[(amminechlorozinc)-μ-chloro]

IP-8.3

End groups that may be considered ionic are named by the principles of coordination nomenclature in the usual manner. The amount of charge is indicated by an Ewens–Bassett number cited at the end of the complete polymer name.

Examples

1. 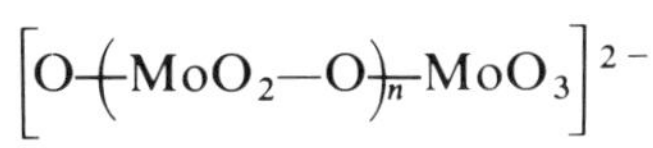

[α-oxo-ω-(trioxomolybdate)-*catena*-poly[(dioxomolybdenum)-μ-oxo]](2 −)

2. 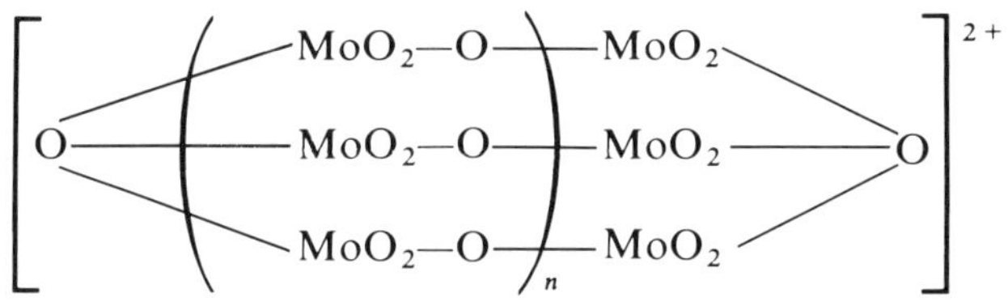

[α-μ_3-oxo-ω-[μ_3-oxo-tris(dioxomolybdenum)]tris[*catena*-poly[(dioxomolybdenum)-μ-oxo]]](2+)*

REFERENCES

1 American Chemical Society. A structure-based nomenclature for linear polymers. *Macromolecules* **1**, 193–198 (1968).

2 IUPAC. Nomenclature of regular single-strand organic polymers: (a) *IUPAC Inf. Bull. Append.* No. **29** (1972); (b) *Macromolecules* **6**, 149–158 (1973) – a separate publication of the tentative rules (Reference [2a]); (c) *Pure Appl. Chem.* **48**, 373–385 (1976) – the approved rules (1975). Reprinted as Chapter 5, this volume.

3 IUPAC. *Nomenclature of organic chemistry*, Sections A, B, C, D, E, F and H, Pergamon Press, Oxford (1979): (a) Rule 6.22, pp. 411–412; (b) Rule 1.1, p. 5.

4 IUPAC. *Nomenclature of inorganic chemistry*, 2nd Edn., Butterworths, London (1971): (a) Rule 7.62, pp. 71–72; (b) Table III, p. 103; (c) Rule 4.14, footnote, p. 27; (d) Table IV, p. 104; (e) Rule 1.4, example 6, p. 12; (f) Rules 7.71 and 7.72, pp. 72–73; (g) Rule 7.62, p. 71.

5 IUPAC. Basic definitions of terms relating to polymers 1974. *Pure Appl. Chem.* **40**, 477–491 (1974): (a) Definition 3.3, p. 482. Reprinted as Chapter 1, this volume.

6 IUPAC. Stereochemical definitions and notations relating to polymers (recommendations 1980). *Pure Appl. Chem.* **53**, 733–752 (1981). Reprinted as Chapter 2, this volume.

*This is a regular single-strand polymer consisting of three chains linked by a terminal oxo end group.

7: Source-Based Nomenclature for Copolymers (1985)

PREAMBLE

Copolymers have gained considerable importance both in scientific research and in industrial applications. A consistent and clearly defined system for naming these polymers would, therefore, be of great utility. The nomenclature proposals presented here are intended to serve this purpose by setting forth a system for designating the types of monomeric-unit sequence arrangements in copolymer molecules.

In principle, a comprehensive structure-based system of naming copolymers would be desirable. However, such a system presupposes a knowledge of the structural identity of all the constitutional units as well as their sequential arrangements within the polymer molecules; this information is rarely available for the synthetic polymers encountered in practice. For this reason, the proposals presented in this Report embody an essentially *source-based nomenclature system.*

Application of this system should not discourage the use of structure-based nomenclature whenever the copolymer structure is fully known and is amenable to treatment by the rules for single-strand polymers [1, 2]. Further, an attempt has been made to maintain consistency, as far as possible, with the abbreviated nomenclature of synthetic polypeptides published by the IUPAC-IUB Commission on Biochemical Nomenclature [3]. It is intended that the present nomenclature system supersede the previous recommendations published in 1952 [4].

BASIC CONCEPT

The nomenclature system presented here is designed for copolymers. By definition, copolymers are polymers that are derived from more than one species of monomer [5]. Various classes of copolymers are discussed, which are based on the characteristic sequence arrangements of the monomeric units within the copolymer molecules. Generally, the names of monomers are used to specify monomeric units; the latter can be named using the trivial, semi-systematic or systematic form. The classes of copolymers are as follows:

unspecified	Rule 1.1
statistical	Rule 2.1
random	Rule 2.2
alternating	Rule 3.1
periodic	Rule 4.1
block	Rule 5.1
graft	Rule 6.1

Prepared by a working group consisting of W. Ring (FRG), I. Mita (Japan), A.D. Jenkins (UK) and N.M. Bikales (USA).
Reprinted from *Pure Appl. Chem.* **57**, 1427–1440 (1985).

In those cases where copolymer molecules can be described by only one species of constitutional unit in a single sequential arrangement, copolymers are regular polymers [5] and can, therefore, be named on a structure basis [1, 2]. Examples will be quoted later in the text.

Polymers having monomeric units differing in constitutional or configurational features, but derived from a single monomer, are not regarded as copolymers, in accordance with the basic definitions [5]. Examples of such polymers, which are not copolymers, are:
(1) polybutadiene with mixed sequences of 1, 2- and 1, 4-units; (2) poly(methyloxirane), also known as poly(propylene oxide), obtained through polymerization of a mixture of the two enantiomers, *R* and *S*, and containing both *R*- and *S*-units.

The nomenclature system presented here can, however, also be applied to such pseudo-copolymers. Polymers having monomeric units differing in constitutional features, but derived from a homopolymer by chemical modification, can be named in the same way, e.g.:
(3) partially hydrolysed poly(vinyl acetate) containing both ester and alcohol units.

A closely related alternative system of nomenclature, which may be preferred in some circumstances, is described in the Appendix.

CLASSIFICATION AND DEFINITION OF COPOLYMERS

A systematic source-based nomenclature for copolymers must identify the constituent monomers and provide a description of the sequence arrangement of the different types of monomeric units present. According to the present proposals, these objectives are achieved by citing the names of the constituent monomers after the prefix 'poly', and by placing between the names of each pair of monomers an italicized connective to denote the kind of arrangement by which those two types of monomeric units are related in the structure. Seven types of sequence arrangement are listed below, together with the corresponding connectives and examples, in which A, B and C represent the names of monomers.

Type	*Connective*	*Example*
unspecified	-*co*-	poly(A-*co*-B)
statistical	-*stat*-	poly(A-*stat*-B)
random	-*ran*-	poly(A-*ran*-B)
alternating	-*alt*-	poly(A-*alt*-B)
periodic	-*per*-	poly(A-*per*-B-*per*-C)
block	-*block*-	polyA-*block*-polyB
graft	-*graft*-	polyA-*graft*-polyB

Each of these types of copolymer is considered in more detail below. When the chemical nature of the end groups is to be specified, the name of the copolymer (as described above) is preceded by the systematic names of the terminal units. The prefix α or ω refers to the terminal unit attached to the left or right, respectively, of the structure, as written.

Example

α-X-ω-Y-poly(A-*co*-B).

The citation of A, B and C is not intended to reflect an order of seniority, unless such seniority is specified in the rules. As a result, more than one name is often possible.

1 Copolymers with an unspecified arrangement of monomeric units

Rule 1.1
An unspecified sequence arrangement of monomeric units is represented by

(A-*co*-B)

and the corresponding copolymer has the name

poly(A-*co*-B).

Example
An unspecified copolymer of styrene and methyl methacrylate is named

poly[styrene-*co*-(methyl methacrylate)].

2 Statistical copolymers

Statistical copolymers are copolymers in which the sequential distribution of the monomeric units obeys known statistical laws; e.g. the monomeric-unit sequence distribution may follow Markovian statistics of zeroth (Bernoullian), first, second or a higher order. Kinetically, the elementary processes leading to the formation of a statistical sequence of monomeric units do not necessarily proceed with equal *a priori* probability. These processes can lead to various types of sequence distribution comprising those in which the arrangement of monomeric units tends towards alternation, tends towards clustering of like units, or exhibits no ordering tendency at all [6]. In simple binary copolymerization, the nature of this sequence distribution can be indicated by the numerical value of a function either of the reactivity ratios or of the related run number [6, 7].

The term statistical copolymer is proposed here to embrace a large proportion of those copolymers that are prepared by simultaneous polymerization of two or more monomers in admixture. Such copolymers are often described in the literature as 'random copolymers', but this is almost always an improper use of the term random and such practice should be abandoned.

Rule 2.1
A statistical sequence arrangement of monomeric units is represented by

(A-*stat*-B), (A-*stat*-B-*stat*-C), etc

where -*stat*- indicates that the statistical sequence distribution with regard to A, B, C, etc, units is considered to be known. Statistical copolymers are named

poly(A-*stat*-B), poly(A-*stat*-B-*stat*-C), etc

Examples

poly(styrene-*stat*-butadiene)

poly(styrene-*stat*-acrylonitrile-*stat*-butadiene)

Random copolymers. A random copolymer is a special case of a statistical copolymer. It is a statistical copolymer in which the probability of finding a given monomeric unit at any given site in the chain is independent of the nature of the neighbouring units at that position (Bernoullian distribution). In other words, for such a copolymer, the probability of finding a sequence ...ABC... of monomeric units A, B, C..., i.e. $P[\ldots\mathrm{ABC}\ldots]$, is given by

$$P[\ldots\mathrm{ABC}\ldots] = P[\mathrm{A}]\cdot P[\mathrm{B}]\cdot P[\mathrm{C}]\ldots = \prod_{i=\mathrm{A,B,C}\ldots} P[i]$$

where *P*[A], *P*[B], *P*[C], etc are the unconditional probabilities of the occurrence of various monomeric units. As already noted above, the term 'random' should *not* be used for statistical copolymers except in this narrow sense.

Some authors use the term 'random' to denote the Bernoullian case further restricted by the condition that the monomeric units be present in exactly equal amounts [8].

Rule 2.2
A random sequence arrangement of monomeric units is represented by

(A-*ran*-B), (A-*ran*-B-*ran*-C), etc

where -*ran*- indicates a random sequence distribution with regard to A, B, C, etc, units. Random copolymers are named

poly(A-*ran*-B), poly(A-*ran*-B-*ran*-C), etc

Example

poly[ethylene-*ran*-(vinyl acetate)]

3 Alternating copolymers

An alternating copolymer is a copolymer comprising two species of monomeric units distributed in alternating sequence. The arrangement

-ABABABAB- or $(AB)_n$

thus represents an alternating copolymer.

Rule 3.1
An alternating sequence arrangement of monomeric units is represented by

(A-*alt*-B)

and the corresponding alternating copolymer is named

poly(A-*alt*-B).

Example

poly[styrene-*alt*-(maleic anhydride)]

Alternating sequence arrangements can form constitutionally regular structures and may, in those cases, also be named utilizing the structure-based nomenclature for regular single-strand organic polymers. The example above would be then named

poly[(2,5-dioxotetrahydrofuran-3,4-diyl)(1-phenylethylene)]

4 Other types of periodic copolymers

In addition to alternating polymers, other structures are known in which the monomeric units appear in an ordered sequence. Examples are:

-ABCABCABC-	or	$(ABC)_n$
-ABBABBABB-	or	$(ABB)_n$
-AABBAABBAABB-	or	$(AABB)_n$
-ABACABACABAC-	or	$(ABAC)_n$

Rule 4.1
A periodic sequence arrangement of monomeric units is represented by

(A-*per*-B-*per*-C)
(A-*per*-B-*per*-B)
(A-*per*-A-*per*-B-*per*-B)
(A-*per*-B-*per*-A-*per*-C), etc

and the corresponding periodic copolymers are named

poly(A-*per*-B-*per*-C)
poly(A-*per*-B-*per*-B)
poly(A-*per*-A-*per*-B-*per*-B)
poly(A-*per*-B-*per*-A-*per*-C), etc, respectively.

If these polymers are regular, they can also be named according to the structure-based nomenclature for regular single-strand organic polymers [1].

Example

The binary monomer mixture consisting of formaldehyde and ethylene oxide might yield the periodically sequenced copolymer

$-\!\!(CH_2{-}O{-}CH_2{-}CH_2{-}O{-}CH_2{-}CH_2{-}O)\!\!-_n$

which is named

poly[formaldehyde-*per*-(ethylene oxide)-*per*-(ethylene oxide)]

or

poly[formaldehyde-*alt*-bis(ethylene oxide)]

or, alternatively,

poly(oxymethyleneoxyethyleneoxyethylene).

Rule 4.2
If copolymer structures comprise several types of periodic sites, only some of which are always occupied by particular species of monomeric units (A, B...), and sites of the other types are occupied by two or more types of monomeric unit (U, V...) in irregular arrangement, the names of the monomers in the latter sites are embraced by parentheses and are separated by semicolon(s).

Examples
1. The copolymer with the sequence arrangement

-AUAVAVAUAVAUAU- is named

poly[A-*alt*-(U; V)].

2. The copolymer with the sequence arrangement

-AUBUAVBUAVBVAUBVAUBU- is named

poly[A-*per*-(U; V)-*per*-B-*per*-(U; V)].

5 Block copolymers
A block polymer is a polymer comprising molecules in which there is a linear arrangement of blocks, a block being defined as a portion of a polymer molecule in which the monomeric units

have at least one constitutional or configurational feature absent from the adjacent portions [5]. In a block copolymer, the distinguishing feature is constitutional, i.e. each of the blocks comprises units derived from a characteristic species of monomer.

In the sequence arrangements

-AAAAAAAA-BBBBBBBBBBB-
-AAAAAAAA-BBBBBBBBBBB-AAAAAAAA-
-AABABAAABB-AAAAAAAA-BBBBBBBBBBB-

the sequences -AAAAAAAA-, -BBBBBBBBBBB-, and -AABABAAABB- are blocks.

Rule 5.1
A block sequence arrangement is represented by

A_k-*block*-B_m, A_k-*block*-(A-*stat*-B), etc

and the corresponding polymers are named

polyA-*block*-polyB, polyA-*block*-poly(A-*stat*-B), etc,

respectively. If no ambiguity arises, a long dash may be used to designate block connections, as follows:

polyA—polyB.

For complex cases, use of -*block*- rather than the long dash is always encouraged. The order of citation of the block names corresponds to the order of succession of the blocks in the chain as written from left to right.

Examples
In the following examples, the subscripts *k*, *m*,...represent different multiplicity of the monomeric units for different blocks. They may be indeterminate or specific (see Rule 5.3). In each case, the first line gives a representation of the block sequence arrangement, the second the corresponding name and the third an illustration of a specific case.

A_k-*block*-B_m
polyA-*block*-polyB
polystyrene-*block*-polybutadiene

A_k-*block*-B_m-*block*-A_k
polyA-*block*-polyB-*block*-polyA
polystyrene-*block*-polybutadiene-*block*-polystyrene

(A-*stat*-B)-*block*-A_k-*block*-B_m
poly(A-*stat*-B)-*block*-polyA-*block*-polyB
poly(styrene-*stat*-butadiene)-*block*-polystyrene-*block*-polybutadiene

A_k-*block*-B_n-*block*-C_m
polyA-*block*-polyB-*block*-polyC
polystyrene-*block*-polybutadiene-*block*-poly(methyl methacrylate).

Rule 5.2
Where a succession of blocks, such as

-A_k-B_n-C_m-

is repeated, the appropriate multiplying prefix is used.

Examples

$(A_k$-*block*-B_n-*block*-$C_m)_3$
tris(polyA-*block*-polyB-*block*-polyC)

$(A_k$-*block*-B_n-*block*-$C_m)_p$
poly(polyA-*block*-polyB-*block*-polyC).

Rule 5.3
When it is possible to specify the chain length of a block, the appropriate Greek prefix (e.g. hecta for 100) may be used rather than poly. Although short sequence lengths are not strictly embraced within the definition of 'block', the same device may usefully be employed by using the general prefix 'oligo' or the appropriate specific prefix (e.g. tri).

Examples

A_c-*block*-B_8
oligoA-*block*-octaB
$(A_c$-*block*-B_k-*block*-$C_3)_n$
poly(oligoA-*block*-polyB-*block*-triC)
$(A_c$-*block*-$B_k)_4$
tetrakis(oligoA-*block*-polyB)

where c is a small integer corresponding to the degree of polymerization of the oligomeric sequence.

Rule 5.4
Those block copolymers, derived from more than two monomers, that also exhibit statistical block sequence arrangements are named according to the principles of Rule 2.1.

Example
The statistical sequence arrangement

-A_k-*block*-B_m-*block*-C_n-*block*-B_m-*block*-A_k-*block*-C_n-

is named

poly(polyA-*stat*-polyB-*stat*-polyC).

Rule 5.5
In the name of block copolymers with blocks connected by way of junction units, X, that are not part of the blocks, the name of the junction unit is inserted in the appropriate place. The connective, -*block*-, may be omitted. Thus,

A_k-*block*-X-*block*-C_m or A_k—X—C_m

is named

polyA-*block*-X-*block*-polyC or polyA—X—polyC.

The same designations can be applied to block polymers.

Examples

polystyrene-*block*-dimethylsilylene-*block*-polybutadiene
or
polystyrene—dimethylsilylene—polybutadiene

polystyrene-*block*-dimethylsilylene-*block*-polystyrene
or
polystyrene—dimethylsilylene—polystyrene.

Rule 5.6
A block copolymer wherein A_k and B_m blocks, connected through junctions X, are distributed in statistical manner in the polymer molecules, as in

-A_k-*block*-X-*block*-B_m-*block*-X-*block*-B_m-*block*-X-*block*-A_k-

is named

poly[(polyA-*block*-X)-*stat*-(polyB-*block*-X)].

A block copolymer wherein A_k and B_m blocks and junction units X are all distributed in statistical manner, as in

-A_k-*block*-X-*block*-B_m-*block*-A_k-*block*-B_m-*block*-X-*block*-B_m-*block*-X-*block*-A_k-

is named

poly(polyA-*stat*-X-*stat*-polyB).

6 Graft copolymers
A graft polymer is a polymer comprising molecules with one or more species of block connected to the main chain as side chains, these side chains having constitutional or configurational features that differ from those in the main chain [5]. In a graft copolymer the distinguishing feature of the side chains is constitutional, i.e. the side chains comprise units derived from at least one species of monomer different from those which supply the units of the main chain.

Rule 6.1
The simplest case of a graft copolymer can be represented by A_k-*graft*-B_m or the arrangement

```
—AAAAAAAAAAAAAAAA—
        |
        B
        B
        B
        B
        B
        B
        B
        |
```

and the corresponding name is

polyA-*graft*-polyB

where the monomer named first (A in this case) is that which supplied the backbone (main chain) units, while that named second (B) is in the side chain(s).

Examples
Each of the following examples presents in order, a representation of the graft sequence arrangement, the corresponding name and an illustration of a specific case.

1. A_k-*graft*-B_m

polyA-*graft*-polyB

polybutadiene-*graft*-polystyrene
(polystyrene grafted to polybutadiene)

2. (A_k-*block*-B_m)-*graft*-C_n

(polyA-*block*-polyB)-*graft*-polyC

(polybutadiene-*block*-polystyrene)-*graft*-polyacrylonitrile
(polyacrylonitrile grafted to a polybutadiene–polystyrene block copolymer at unspecified sites)

3. (A-*stat*-B)-*graft*-C_n

poly(A-*stat*-B)-*graft*-polyC

poly(butadiene-*stat*-styrene)-*graft*-polyacrylonitrile
(polyacrylonitrile grafted to a statistical butadiene–styrene copolymer at unspecified sites)

4. A_k-*block*-(B_m-*graft*-C_n)

polyA-*block*-(polyB-*graft*-polyC) or (polyB-*graft*-polyC)-*block*-polyA

polybutadiene-*block*-(polystyrene-*graft*-polyacrylonitrile) or
(polystyrene-*graft*-polyacrylonitrile)-*block*-polybutadiene
(polystyrene–polybutadiene block copolymer with polyacrylonitrile grafted to the polystyrene block)

5. [A_k-*graft*-(B-*co*-C)]-*block*-B_m

[polyA-*graft*-poly(B-*co*-C)]-*block*-polyB or polyB-*block*-[polyA-*graft*-poly(B-*co*-C)]

[polybutadiene-*graft*-poly(styrene-*co*-acrylonitrile)]-*block*-polystyrene or
polystyrene-*block*-[polybutadiene-*graft*-poly(styrene-*co*-acrylonitrile)]
(polybutadiene–polystyrene block copolymer with a styrene–acrylonitrile copolymer with an unspecified sequence arrangement of monomeric units grafted to the polybutadiene block).

Rule 6.2
If more than one type of graft chain is attached to the backbone, semicolons are used to separate the names of the grafts or their symbolic representations.

Example

A_k-*graft*-(B_m; C_n)

polyA-*graft*-(polyB; polyC)

polybutadiene-*graft*-[polystyrene; poly(methyl methacrylate)]
(polystyrene and poly(methyl methacrylate) chains grafted to a polybutadiene backbone).

Rule 6.3
Graft copolymers with known numbers of graft chains are named using numeric prefixes (mono, bis, tris, etc).

Example

A_k(-*graft*-B_m)$_3$

polyA-tris(-*graft*-polyB)

polybutadiene-tris(-*graft*-polystyrene)
(three polystyrene grafts per polybutadiene backbone).

If the precise site of grafting is known, it can be specified.

Example

A_{10}-*block*-(X-*graft*-B_m)-*block*-A_{15}

decaA-*block*-(X-*graft*-polyB)-*block*-pentadecaA

decabutadiene-*block*-(methylsilanetriyl-*graft*-polystyrene)-*block*-pentadecabutadiene.

The system of naming graft copolymers is also applicable, in principle, to 'star copolymers', where chains having different constitutional or configurational features are linked through a central moiety.

Examples

1. A_k-*block*-[X-(*graft*-B_m)$_2$]-*block*-A_k or B_m-*block*-[X-(*graft*-A_k)$_2$]-*block*-B_m

polyA-*block*-[X-bis(-*graft*-polyB)]-*block*-polyA or
polyB-*block*-[X-bis(-*graft*-polyA)]-*block*-polyB

polystyrene-*block*-[silanetetrayl-bis(-*graft*-polybutadiene)]-*block*-polystyrene or
polybutadiene-*block*-[silanetetrayl-bis(-*graft*-polystyrene)]-*block*-polybutadiene
(two polystyrene and two polybutadiene chains attached to a central Si atom)

2. A_k-*block*-[X-*graft*-(B_m; C_n)]-*block*-D_p

polyA-*block*-[X-*graft*-(polyB; polyC)]-*block*-polyD

polystyrene-*block*-[silanetetrayl-*graft*-(polybutadiene; polyisoprene)]-*block*-poly(methyl methacrylate)
(a polybutadiene chain, a polyisoprene chain, a polystyrene chain and a poly(methyl methacrylate) chain attached to the same central Si atom).

In the absence of a seniority rule, several other names are possible.

7 Polymers made by condensation polymerization or related polymerization

The nomenclature system for copolymers is also applicable to polymers made by condensation polymerization of more than one monomeric species, or, more generally, by polymerization of more than one monomeric species where molecules of all sizes (i.e. monomers, oligomers, polymers) can react with each other. One can distinguish the case of polymers made by condensation polymerization of homopolymerizable monomers from that of polymers made by condensation polymerization of complementary ingredients that do not usually separately homopolymerize.

Rigorous application of the source-based definition of a copolymer [5] embraces polymers such as poly(ethylene terephthalate) or poly(hexamethylene adipamide) (which are commonly regarded as homopolymers) because two ingredients are, in each case, the usual starting materials of polymerization. If polymers of this type have constitutionally regular structures and are regular polymers, the nomenclature for regular single-strand organic polymers can also be used [1].

This applies, for example, to the polymer derived from terephthalic acid and ethylene glycol, which by source-based copolymer nomenclature would be named as poly[(ethylene glycol)-*alt*-(terephthalic acid)], if in fact the polymer has been prepared by a condensation polymerization starting with terephthalic acid and ethylene glycol. However, if the starting material is the partial ester, $HOCH_2CH_2OCOC_6H_4COOH$, the appropriate source-based name is that of a homopolymer, whereas use of the starting material bis(2-hydroxyethyl) terephthalate, $HOCH_2CH_2OCOC_6H_4COOCH_2CH_2OH$ (extensively employed industrially), would suggest the name poly[bis(2-hydroxyethyl) terephthalate]. Regardless of the starting materials used, the structure-based name is poly(oxyethyleneoxyterephthaloyl). The trivial name poly(ethylene terephthalate) is also permitted, because it is so well established in the literature.

For all such polymers made by condensation polymerization of two complementary difunctional ingredients (or 'monomers'), which can readily be visualized as reacting on a 1:1 basis to give an 'implicit monomer', the homopolymerization of which would give the actual product, the single-strand structure-based nomenclature may be suitable insofar as such a polymer is regular and can be represented as possessing a single constitutional repeating unit. It is to be noted that this is applicable only to cases where the mole ratio of the ingredients is 1:1 and the ingredients are exclusively difunctional.

The introduction of a third component into the reaction system necessitates the use of copolymer nomenclature which can logically be developed from the foregoing rules, as the examples below illustrate. The copolymer derived from reaction of ethylene glycol with a mixture of terephthalic and isophthalic acids would be named:

poly[[(ethylene glycol)-*alt*-(terephthalic acid)]-*co*-[(ethylene glycol)-*alt*-(isophthalic acid)]]

poly[(ethylene terephthalate)-*co*-(ethylene isophthalate)]

or

poly[(ethylene glycol)-*alt*-(terephthalic acid; isophthalic acid)]

A copolymer formed from oligo[(adipic acid)-*alt*-(1,4-butanediol)] and oligo[(2,4-tolylene diisocyanate)-*co*-(trimethylolpropane)] in presence of trimethylolpropane is named poly[oligo-[(adipic acid)-*alt*-(1,4-butanediol)]-*co*-oligo[(2,4-tolylene diisocyanate)-*co*-(trimethylolpropane)]-*co*-(trimethylolpropane)]

A polymer derived from the condensation polymerization of a single actual monomer, the molecules of which terminate in two different complementary functional groups (e.g. 6-aminohexanoic acid) is, by definition, a (regular) homopolymer. When two different monomers of this type react together, the product is a copolymer that can be named in appropriate fashion. For example, if 6-aminohexanoic acid is copolycondensed with 7-aminoheptanoic acid, leading to a statistical distribution of monomeric units, the product is named poly[(6-aminohexanoic acid)-*stat*-(7-aminoheptanoic acid)].

8 Specification with regard to mass fractions, mole fractions, molar masses and degrees of polymerization

Whereas subscripts placed immediately after the formula of the monomeric unit or the block designate the degree of polymerization or repetition, mass and mole fractions and molar masses—which in most cases are average quantities—may be expressed by placing corresponding figures after the complete name or symbol of the copolymer. The order of citation is the same as for the monomer species in the name or symbol of the copolymer. Unknown quantities can be designated by *a*, *b*, etc.

Although this scheme can be extended to complicated cases, it is recommended that its application be restricted to simple cases; any treatment of more complicated systems should be explained in the text.

Rule 8.1

Mass fractions, or mass percentages, for the monomeric units are placed in parentheses after the copolymer name, followed by the symbol 'w', or the phrase 'mass %', respectively. The order of citation in the parentheses is the same as in the name.

Examples

1. polybutadiene-*graft*-polystyrene (0.75:0.25 w) or polybutadiene-*graft*-polystyrene (75:25 mass %) (a graft copolymer containing 75 mass % of polybutadiene and 25 mass % of grafted polystyrene)

2. polybutadiene-*graft*-poly(styrene-*stat*-acrylonitrile) (0.75:a:b w) or polybutadiene-*graft*-poly(styrene-*stat*-acrylonitrile) (75:a:b mass %) (a graft copolymer containing 75 mass % of butadiene units as backbone and unknown quantities in statistical arrangement of styrene and acrylonitrile units in the grafted chains).

Rule 8.2

Mole fractions, or mole percentages, for the monomeric units are placed in parentheses after the copolymer name, followed by the symbol 'x', or the phrase 'mol %', respectively. The order of citation in the parentheses is the same as in the name.

Example

polybutadiene-*graft*-polystyrene (0.85:0.15 x) or
polybutadiene-*graft*-polystyrene (85:15 mol %)
(a graft copolymer containing 85 mol % of butadiene units and 15 mol % of styrene units).

Rule 8.3

The molar mass, relative molecular mass or degree of polymerization may be included in the scheme of Rules 8.1 and 8.2 by adding the corresponding figures, followed by the symbol M, M_r or DP, respectively.

Examples

1. polybutadiene-*graft*-polystyrene (75:25 mass %; 90 000:30 000 M_r) (a graft copolymer consisting of 75 mass % of butadiene units with a relative molecular mass of 90 000 as the backbone, and 25 mass % of styrene units in grafted chains with a relative molecular mass of 30 000)

2. polybutadiene-*graft*-polystyrene (1700:290 DP) (a graft copolymer consisting of a polybutadiene backbone with a degree of polymerization of 1700 to which polystyrene with a degree of polymerization of 290 is grafted).

APPENDIX: ALTERNATIVE NOMENCLATURE FOR COPOLYMERS

The nomenclature system for copolymers developed in the main text of this Report is designed to meet the requirement of providing a systematic name for any copolymer, however complex the structure; as a result, the systematic names may be undesirably long in some cases. However, many copolymers reported in the literature have relatively simple structures, which do not necessitate this elaborate system. For these simpler cases, a different, more concise nomenclature system is presented here as an alternative.

Fundamental principles

The alternative nomenclature system is based on the following principles.

1. A copolymer is described by the prefix 'copoly' followed by citation of the names of the monomers used (source-based nomenclature). The prefix is used only once, just as the prefix 'poly' is employed only once in naming regular single-strand polymers (called regular polymers hereafter) [1] or the simple copolymers described in the main text.

Example

copoly(styrene/butadiene).

2. The specification of the type of structure in a copolymer (the connectives *block*, *alt*, etc of the main nomenclature system) is shown by an italicized prefix preceding 'copoly.'

Example

block-copoly(styrene/butadiene).

3. Only the names of the monomers are included in the main part of the copolymer name; the terminal units are specified before the main name (using the prefixes α and ω), whereas the junction units between blocks are specified after the main name (using the symbol μ).

Example

block-copoly(styrene/butadiene)-μ-dimethylsilylene.

4. The mass fraction, mole fraction, molar mass or degree of polymerization of monomeric units in copolymers is specified separately in parentheses after the name.

5. In addition to the principles stated in (1)–(4), above, some further conventions are used for the more complex copolymers, e.g. where the structure cannot be classified in a unique fashion.

In general, no seniority rule is provided for the order of citation of the monomer names. In block or periodic copolymers, the order of citation of the names of monomers corresponds to the sequence in which the monomeric units occur in the molecules. In graft copolymers, the initially cited name is that of the backbone (main chain).

A1 Application to simple copolymers

In the examples cited below, the name derived from the system elaborated in the main text of this Report is given after the alternative name proposed in this Appendix.

Rule A1.1

The name of a copolymer consists of the prefix 'copoly' followed by parentheses in which the names of the monomers used are enumerated, separated by an oblique stroke, as in copoly(A/B), copoly(A/B/C), etc, where A, B and C represent the names of the monomers employed.

Example

copoly(styrene/methyl methacrylate) alternative name
poly[styrene-*co*-(methyl methacrylate)] name from main text

Rule A1.2

The arrangement of monomeric units in a copolymer, if known, is specified by one of the following italicized prefixes: *stat*- (statistical), *ran*- (random), *alt*- (alternating), *per*- (periodic), *block*- or *graft*- (see 'Classification and definition of copolymers', p. 131).

Examples

stat-copoly(styrene/butadiene)
poly(styrene-*stat*-butadiene)

ran-copoly(ethylene/vinyl acetate)
poly[ethylene-*ran*-(vinyl acetate)]

alt-copoly(styrene/maleic anhydride)
poly[styrene-*alt*-(maleic anhydride)]

block-copoly(styrene/butadiene/methyl methacrylate)
polystyrene-*block*-polybutadiene-*block*-poly(methyl methacrylate)

graft-copoly(butadiene/styrene)
polybutadiene-*graft*-polystyrene

Rule A1.3
When monomeric units of one particular kind occur in groups in a periodic copolymer, this can be indicated by a 'polykis' prefix. The repetition of a set of blocks in a block copolymer for a known or unknown number of times can be represented similarly.

Examples

per-copoly(A/B/B/B) = *per*-copoly(A/trisB) = *alt*-copoly(A/trisB)
poly(A-*per*-trisB)

block-copoly(A/B/C/A/B/C/A/B/C) = *block*-copoly[tris(A/B/C)]
tris(polyA-*block*-polyB-*block*-polyC)

per-copoly[formaldehyde/bis(ethylene oxide)]
poly[formaldehyde-*per*-bis(ethylene oxide)]

Rule A1.4
When one type of site in a basically alternating or periodic copolymer can be occupied by units derived from two or more monomers (such as in the copolymers considered in Rule 4.2 of the main text), the names of the copolymers are based on the principles of the main text and the rules in this Appendix.

Examples

per-copoly[A/B/(C;D)]
poly[A-*per*-B-*per*(C;D)]

alt-copoly[methyl methacrylate/(styrene; 1-vinylnaphthalene)]
poly[(methyl methacrylate)-*alt*-(styrene; 1-vinylnaphthalene)]

The same procedure can be used to name graft copolymers with two or more different types of of branch (B, C, etc) grafted onto a backbone (A).

Examples

graft-copoly[A/(B;C)]
polyA-*graft*-(polyB; polyC)

graft-copoly[butadiene/(styrene; methyl methacrylate)]
polybutadiene-*graft*-[polystyrene; poly(methyl methacrylate)]

Rule A1.5

Terminal units (preceded by the prefixes α and ω) are specified before the main copolymer name, but junction units between blocks (preceded by the prefix μ) are specified after the main copolymer name. If one type of junction unit occurs in the structure more than once, the appropriate multiplying prefix (bis, tris, etc) may be used; if more than one type of junction unit occurs in the structure, they may be designated μ_1, μ_2, etc.

Examples

α-butyl-ω-carboxy-*block*-copoly(styrene/butadiene)
α-butyl-ω-carboxy-polystyrene-*block*-polybutadiene

block-copoly(styrene/butadiene)-μ-dimethylsilylene
polystyrene-*block*-dimethylsilylene-*block*-polybutadiene

block-copoly(styrene/butadiene/styrene)-bis(μ-dimethylsilylene)
polystyrene-*block*-dimethylsilylene-*block*-polybutadiene-*block*-dimethylsilylene-*block*-polystyrene

graft-copoly(butadiene/styrene)-polykis(μ-methylsilanetriyl).

Rule A1.6

Specification with regard to mass fractions, mole fractions, molar masses, relative molecular masses and degrees of polymerization is treated as in the rules of Section 8 in the main text.

A2 Application to more complex copolymers

The alternative nomenclature can be extended to copolymers of some complexity. The following rules and examples deal with a few such cases. In general, the nomenclature system described in the main part of this report will be found preferable for complex structures.

Rule A2

When a graft copolymer or block copolymer contains a constituent block which is itself a copolymer, the block is named co(B/C) with a descriptive prefix, if necessary (see Rule A1.2).

Examples

block-copoly[*stat*-co(styrene/butadiene)/styrene/butadiene]
poly(styrene-*stat*-butadiene)-*block*-polystyrene-*block*-polybutadiene

block-copoly[styrene/-*graft*-co(butadiene/acrylonitrile)]
polystyrene-*block*-(polybutadiene-*graft*-polyacrylonitrile)

graft-copoly[*stat*-co(butadiene/styrene)/acrylonitrile]
poly(butadiene-*stat*-styrene)-*graft*-polyacrylonitrile

block-copoly[*graft*-co[butadiene/-*stat*-co(styrene/acrylonitrile)]/styrene]
[polybutadiene-*graft*-poly(styrene-*stat*-acrylonitrile)]-*block*-polystyrene

A3 Application to polymers made by condensation polymerization or related polymerization

The principles of the main text and the rules in this Appendix can be applied to polymers made by condensation polymerization or related polymerization.

Examples of alternative names

copoly(6-aminohexanoic acid/7-aminoheptanoic acid)

alt-copoly(ethylene glycol/terephthalic acid)

alt-copoly[ethylene glycol/(terephthalic acid; isophthalic acid)]

REFERENCES

1 IUPAC. Nomenclature of regular single-strand organic polymers (rules approved 1975). *Pure Appl. Chem.* **48**, 373–385 (1976). Reprinted as Chapter 5, this volume.

2 IUPAC. Nomenclature for regular single-strand and quasi-single-strand inorganic and coordination polymers (recommendations 1984). *Pure Appl. Chem.* **57**, 149–168 (1985). Reprinted as Chapter 6, this volume.

3 IUPAC-IUB. Abbreviations and symbols for nucleic acids, polynucleotides and their constituents (recommendations 1970). *Pure Appl. Chem.* **40**, 277–290 (1974).

4 IUPAC. Report on nomenclature in the field of macromolecules. *J. Polym. Sci.* **8**, 257–277 (1952).

5 IUPAC. Basic definitions of terms relating to polymers 1974. *Pure Appl. Chem.* **40**, 477–491 (1974). Reprinted as Chapter 1, this volume.

6 G.E. Ham. In *Encyclopedia of Polymer Science and Technology* (Eds H.F. Mark, N.G. Gaylord, and N.M.). Bikales, Wiley–Interscience, New York, **4**, 165 (1966).

7 H.J. Harwood and W.M. Ritchey. The characterization of sequence distribution in copolymers. *J. Polym. Sci., Polym. Lett. Ed.* **2**, 601 (1964).

8 H.J. Harwood. Characterization of the structures of diene polymers by NMR. *Rubber Chem. Technol.* **55**, 769 (1982).

8: A Classification of Linear Single-Strand Polymers (1988)

PREAMBLE

The polymer literature lacks, at present, a widely applicable, internally consistent, and easily recognizable system for classifying polymers. This situation prevails in spite of the large number of polymers that are already known or will be synthesized in the future. The classification scheme to be presented here facilitates logical indexing, including through the use of computer techniques, information storage and retrieval. At the present time, the scheme is limited to linear single-strand organic and inorganic polymers. It is consistent with previous IUPAC recommendations [1–5].

GENERAL PRINCIPLES

The classification consists of a hierarchical scheme for naming polymers according to the chemical constitution of the repeating units in the main chain (backbone). It can be applied to homopolymers, alternating copolymers and other macromolecular substances in which the constitutional repeating units can be identified. All existing linear single-strand polymers are embraced by this classification, which has been designed so as to be capable of extension to include any new structures of this type. In view of the open-ended nature of the classification, the tables below can only be illustrative, not comprehensive.

Four hierarchical levels are used; in order of decreasing importance, they are: classes, subclasses, groups and individual polymers.

CLASSES

Polymers are divided into two principal classes on the basis of the constitution of the main chain. *Homochain polymers* are those in which the main chains are constructed from atoms of a single element; *heterochain polymers* are those in which the main chains are constructed from atoms of two or more elements.

Rule 1.1
Homochain polymers are named by placing the name or symbol of the element in the main chain immediately before the expression '-chain polymer'.

Examples (see also 'Subclasses', p. 149, and Fig. 8.1)
1. Carbon-chain polymer or C-chain polymer.
2. Sulfur-chain polymer or S-chain polymer.

Prepared by a working group consisting of N.A. Platé (USSR) and I.M. Papisov (USSR).
Reprinted from *Pure Appl. Chem.* **61**, 243–254 (1989).

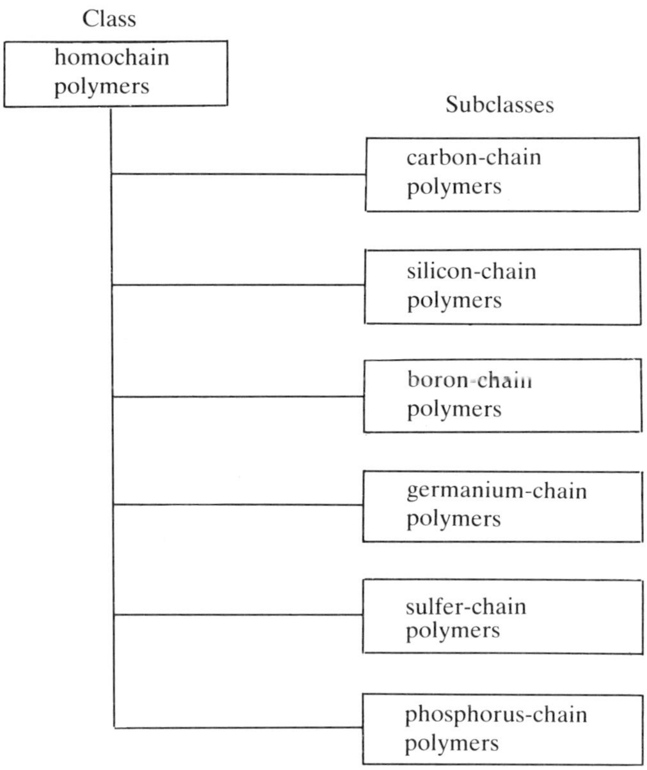

Fig. 8.1 Examples of subclasses of homochain polymers.

Rule 2.1
Heterochain polymers are named by placing the names or symbols of all the elements in the main chain, in parentheses, immediately before the expression '-chain polymer'.

Examples (see also 'Subclasses', p. 149, and Fig. 8.2)
1. (Oxygen, carbon)-chain polymer or (O, C)-chain polymer.
2. (Oxygen, nitrogen, carbon)-chain polymer or (O, N, C)-chain polymer.

Rule 2.2
The order of citation of the elements in heterochain polymers is that conventionally used in inorganic nomenclature [6]. For the common elements, the order of citation is as follows: O, S, N, P, C, Si, B. Note that in those cases where the main chain has bonds in common with cyclic structures, *all* atoms in the rings must be considered for classification purposes. Thus I is a homochain polymer, whereas II is a heterochain polymer.

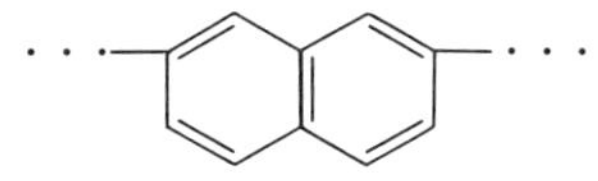
I

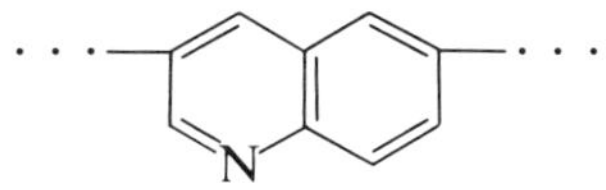
II

On the other hand, a polymer such as III, in which the main chain has no bonds in common with the cyclic structure, is classified as a homochain polymer.

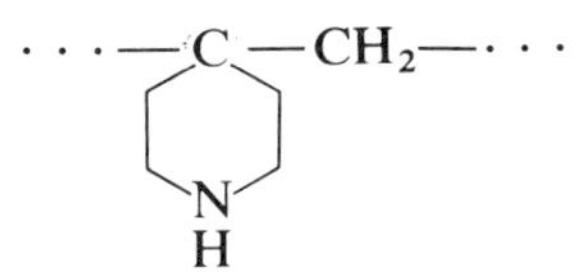
III

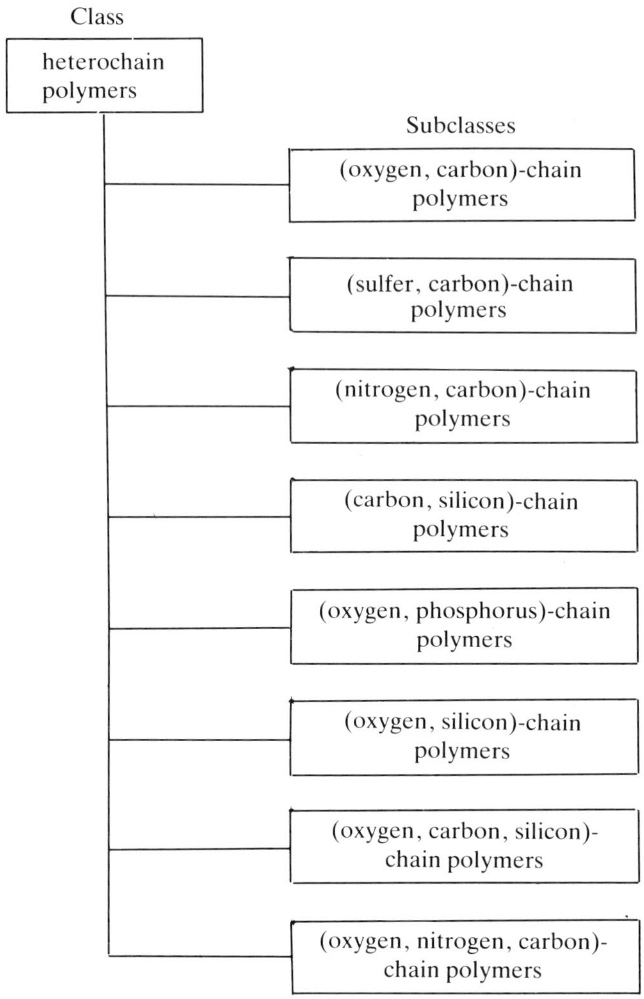

Fig. 8.2 Examples of subclasses of heterochain polymers.

Rule 3.1
The presence of specific side groups, or specific elements in side groups, is indicated by placing the name of the side group, or of the element in the side group, immediately before the expression '-side-group polymer'.

Rule 3.2
Where more than one element, or more than one side group, is to be specified, the names of the elements or side groups are placed in parentheses.

Examples
1. Oxygen-side-group polymer.
2. Hydroxy-side-group polymer [for poly(1-hydroxyethylene), etc].
3. Ether-side-group polymer [for poly(1-methoxyethylene), etc].
4. (O, N, P)-side-group polymer.
5. (Ether, amide)-side-group polymer.

SUBCLASSES

Each class of polymers can be divided into various subclasses, according to the nature of the elements in the main chain. Examples are given in Rules 1.1, 2.1, as well as in Figs 8.1 and 8.2.

GROUPS

Each subclass can be further divided into groups having similar chemical structures. Examples of carbon-chain polymers are polyalkylenes, polyarylenes, etc (Fig. 8.3).

Examples of (oxygen, carbon)-chain polymers are polyethers, polycarbonates, etc (Fig. 8.4). Examples of other types can be found in Figs 8.5 and 8.6.

It is convenient to retain such widely accepted terminology as polyacetals, polycarbonates, polyamides, polyesters, nucleic acids, etc, for naming the groups of polymers.

INDIVIDUAL POLYMERS

The lowest hierarchical ranking is given to the individual polymers, which are named in accordance with accepted IUPAC nomenclature practice. The reader is referred to References [2] and [3] for recommendations for naming organic and inorganic single-strand polymers, respectively.

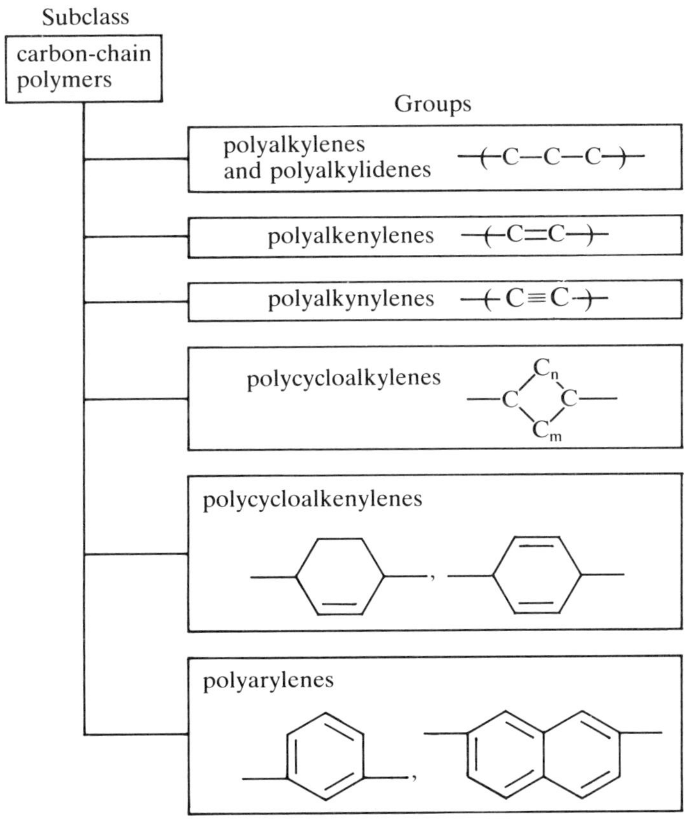

Fig. 8.3 Examples of groups of carbon-chain polymers.

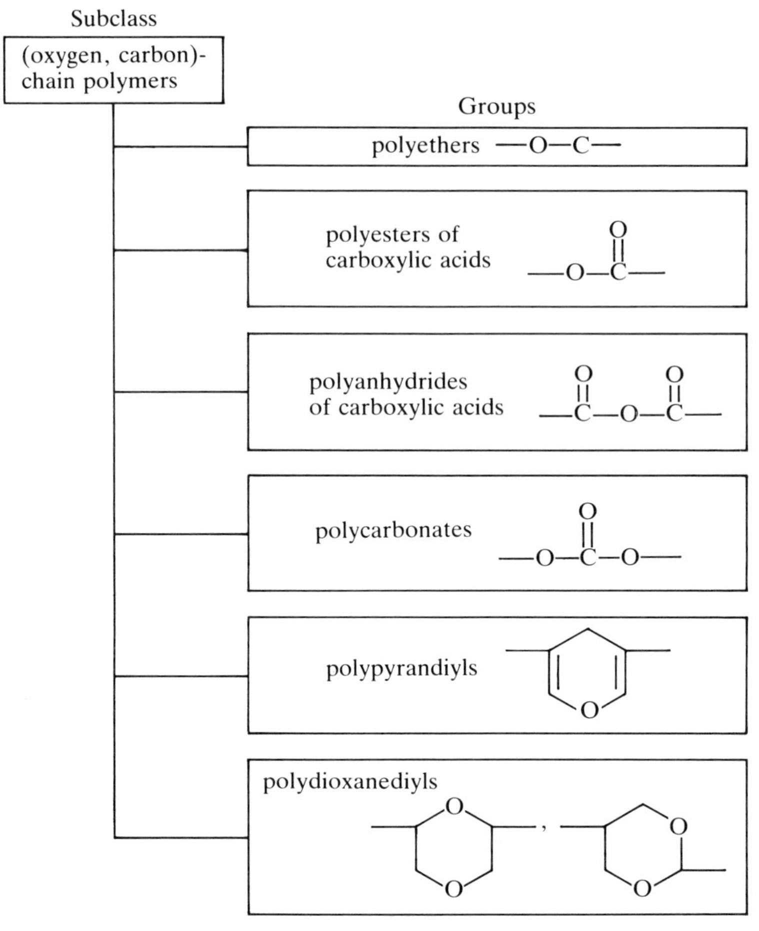

Fig. 8.4 Examples of groups of (oxygen, carbon)-chain polymers.

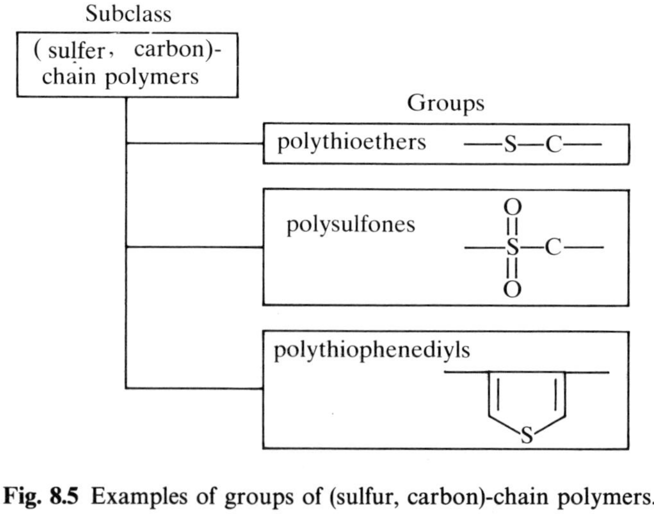

Fig. 8.5 Examples of groups of (sulfur, carbon)-chain polymers.

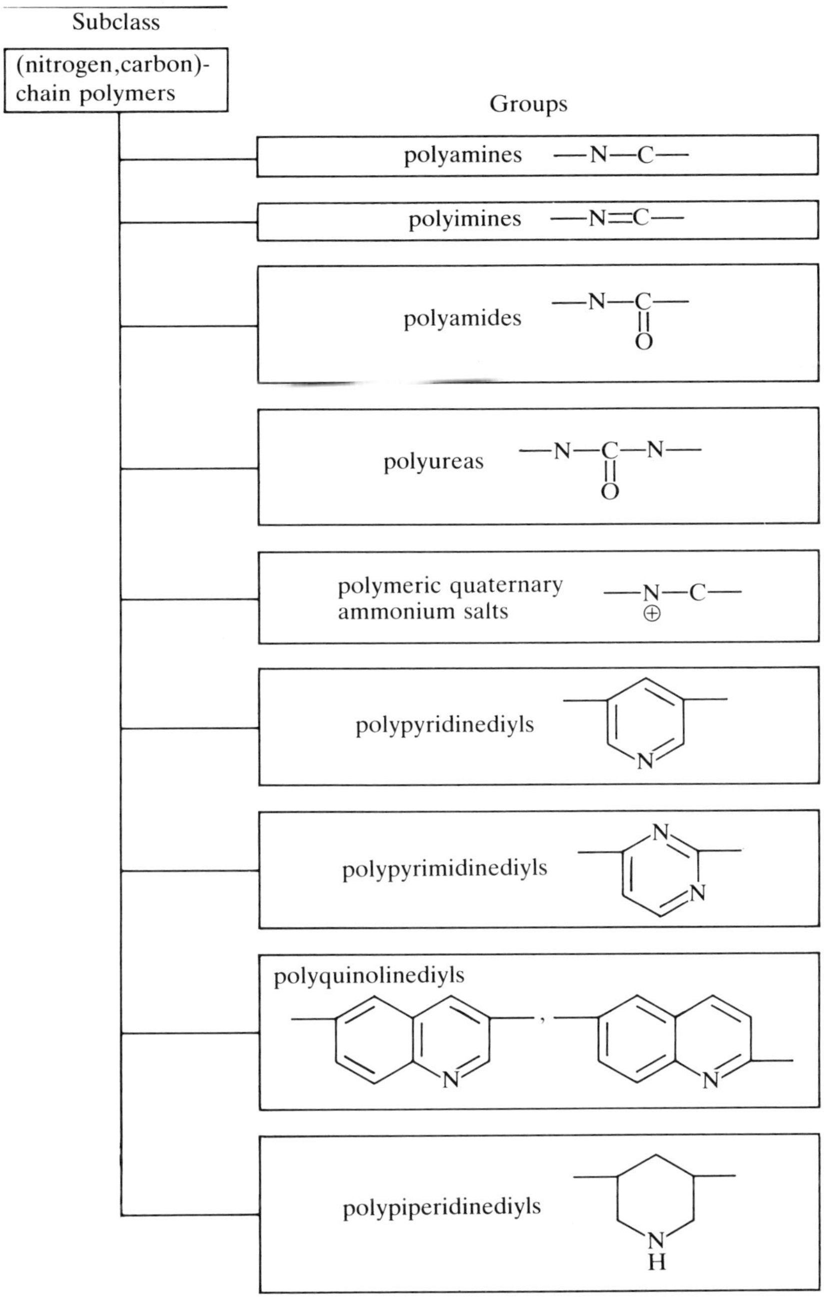

Fig. 8.6 Examples of groups of (nitrogen, carbon)-chain polymers.

It is to be noted that, for polymers of complex structure, assignment to more than one subclass and/or group is sometimes possible. For example, the polymer

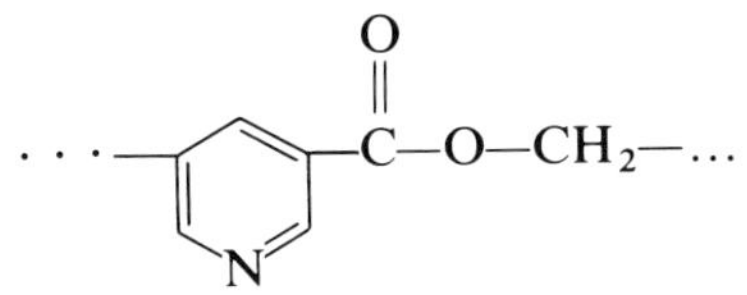

may be classified as follows:

class: heterochain
subclass: (O, N, C)-chain polymer
groups: polycarboxylate and/or pyridine-chain polymer

Additional examples are given in Table 8.1 and the Appendix.

Table 8.1 Examples of individual polymers

Group	Individual polymer	
	Structure	Name
polyalkylenes	$\left(\overset{CH_3}{\underset{CH_3}{C}}CH_2\right)_n$	poly(1,1-dimethylethylene) polyisobutylene
polyalkenylenes	$\left(\underset{Cl}{C}{=}CHCH_2CH_2\right)_n$	poly(1-chloro-l-butenylene) polychloroprene
polyethers	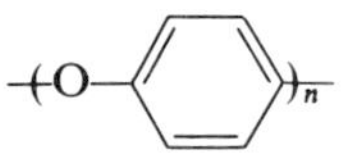	poly(oxy-1,4-phenylene) poly(phenylene oxide)
polyamides	$\left[NH\overset{O}{\overset{\Vert}{C}}(CH_2)_4\overset{O}{\overset{\Vert}{C}}NH(CH_2)_6\right]_n$	poly(iminoadipoyliminohexamethylene) poly(hexamethylene adipamide)

REFERENCES

1 IUPAC. Basic definitions of terms relating to polymers 1974. *Pure Appl. Chem.* **40**, 477–491 (1974). Reprinted as Chapter 1, this volume.
2 IUPAC. Nomenclature of regular single-strand organic polymers (rules approved 1975). *Pure Appl. Chem.* **48**, 373–385 (1976). Reprinted as Chapter 5, this volume.
3 IUPAC. Nomenclature for regular single-strand and quasi-single-strand inorganic and coordination polymers (recommendations 1984). *Pure Appl. Chem.* **57**, 149–168 (1985). Reprinted as Chapter 6, this volume.
4 IUPAC. Stereochemical definitions and notations relating to polymers (recommendations 1980). *Pure Appl. Chem.* **53**, 733–752 (1981). Reprinted as Chapter 2, this volume.
5 IUPAC. Report on nomenclature in the field of macromolecules, *J. Polym. Sci.* **8**, 257–277 (1952).
6 Rule IP-2.1 in Ref. [3].

APPENDIX

CLASSIFICATION OF SINGLE-STRAND POLYMERS

	Individual polymer		Group	Subclass	Class
1.	$-(S)_n-$	*catena*-poly[sulfur]	—	S-chain	homochain
2.		poly(1,4-phenylene)	polyarylenes	C-chain	homochain
3.	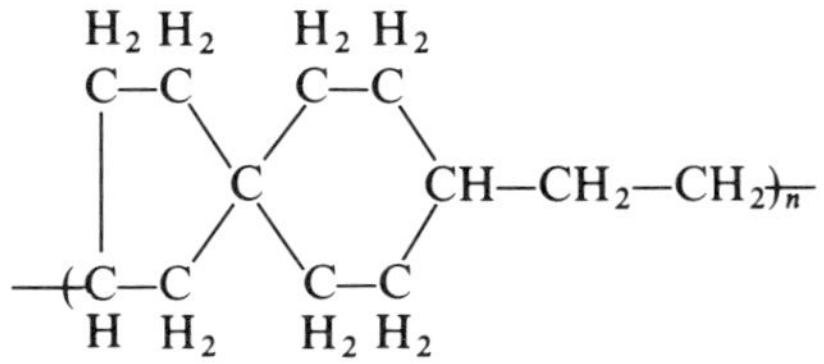	poly(spiro[4.5]dec-2,8-yleneethylene)	polycycloalkylenes	C-chain	homochain
4.	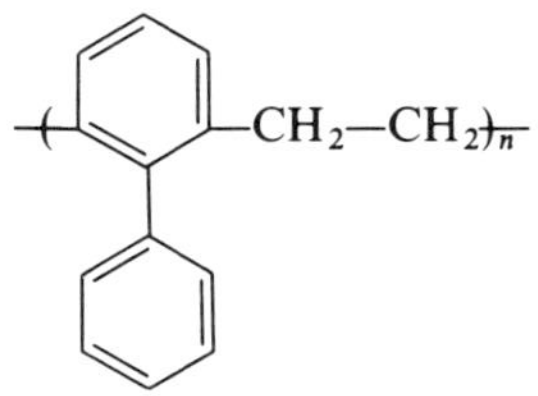	poly(2,6-biphenylyleneethylene)	polyarylenes	C-chain	homochain
5.	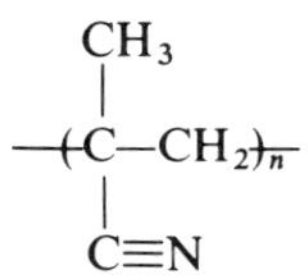	poly(1-cyano-1-methylethylene)	polyalkylenes	C-chain	homochain
6.	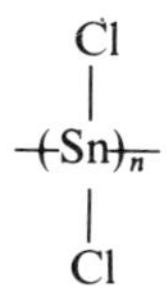	*catena*-poly[dichlorotin]	—	Sn-chain	homochain

Individual polymer		Group	Subclass	Class
7. 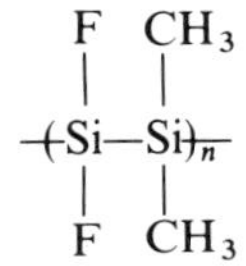	*catena*-poly[(difluorosilicon)(dimethyl-silicon)]	—	Si-chain	homochain
8. 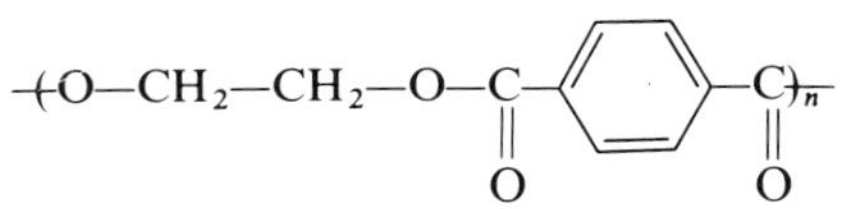	poly(oxyethyleneoxyterephthaloyl)	polyesters of carboxylic acids	(O, C)-chain	heterochain
9. 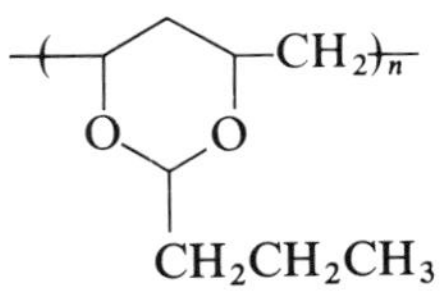	poly[(2-propyl-1,3-dioxane-4,6-diyl)methylene]	polydioxanediyls, polyacetals	(O, C)-chain	heterochain
10. 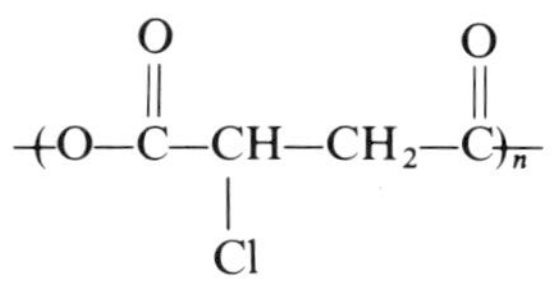	poly[oxy(2-chlorosuccinyl)]	polyanhydrides of carboxylic acids	(O, C)-chain	heterochain
11. 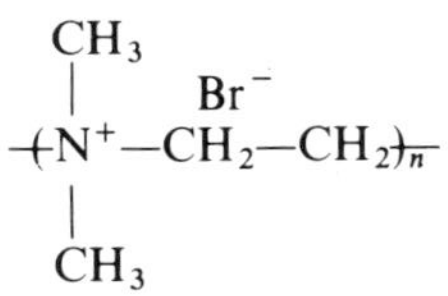	poly[(dimethyliminio)-ethylene bromide]	polymeric quaternary ammonium salts	(N, C)-chain	heterochain
12. 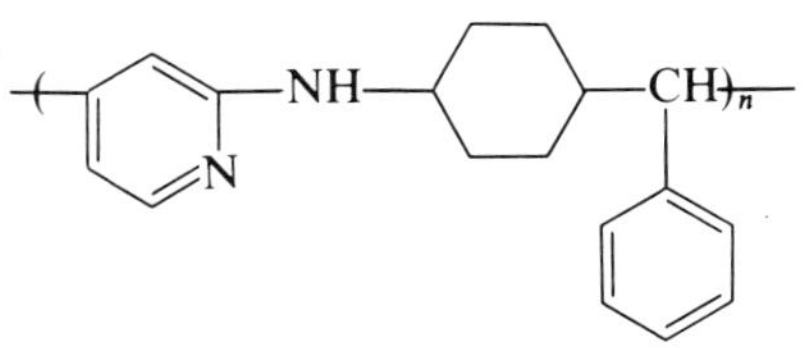	poly(4,2-pyridinediylimino-1,4-cyclohexylenebenzylidene)	polyamines and/or polypyridinediyls	(N, C)-chain	heterochain

	Individual polymer		Group	Subclass	Class
13.	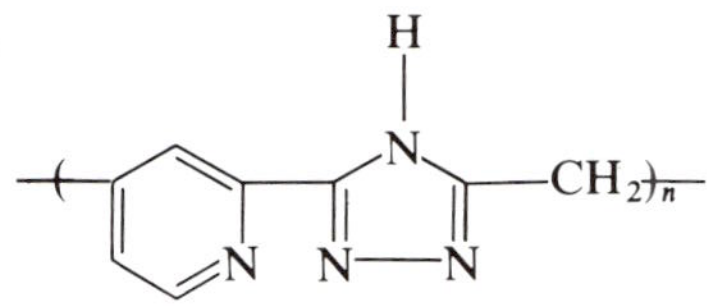	poly(4,2-pyridinediyl-4*H*-1,2,4-triazole-3,5-diylmethylene)	polypyridinediyls and/or polytriazolediyls	(N, C)-chain	heterochain
14.	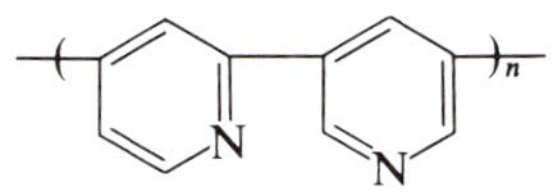	poly([2,3′-bipyridine]-4,5′-diyl)	polypyridinediyls	(N, C)-chain	heterochain
15.	$-[NH-C(=O)-CH_2]_n-$	poly[imino(1-oxoethylene)]	polyamides (or polypeptides)	(N, C)-chain	heterochain
16.	$-[Si(Ph)_2-O]_n-$	*catena*-poly[(diphenylsilicon)-μ-oxo] poly(diphenylsiloxane)	polysiloxanes	(O, Si)-chain	heterochain
17.	$-[Al(H)-N(Ph)]_n-$	*catena*-poly[(hydridoaluminium)-μ-(phenylimido)]	—	(N, Al)-chain	heterochain
18.	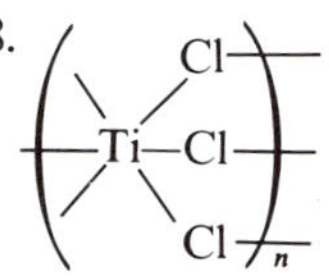	*catena*-poly[titaniumtri-μ-chloro]	—	(Cl, Ti)-chain	heterochain
19.	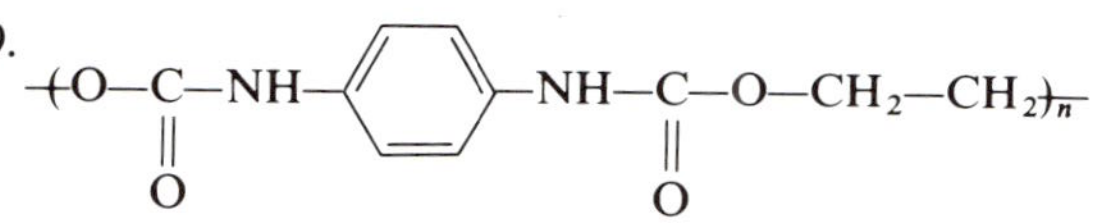	poly(oxycarbonylimino-1,4-phenyleneimino-carbonyloxyethylene)	polyurethanes	(O, N, C)-chain	heterochain

	Individual polymer		Group	Subclass	Class.
20.	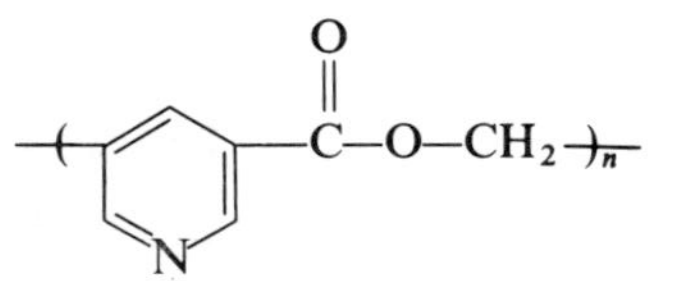	poly(3,5-pyridinediylcarbonyloxymethylene)	polyesters of carboxylic acids and/or polypyridinediyls	(O, N, C)-chain	heterochain
21.	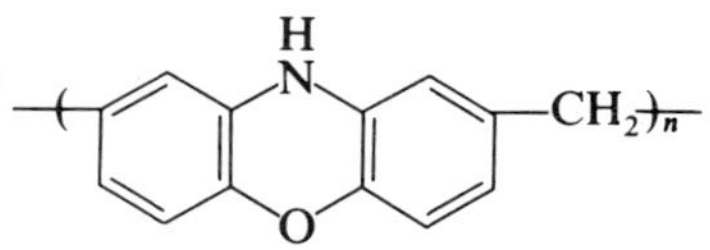	poly(2,8-phenoxazinediylmethylene)	polyphenoxazine	(O, N, C)-chain	heterochain
22.	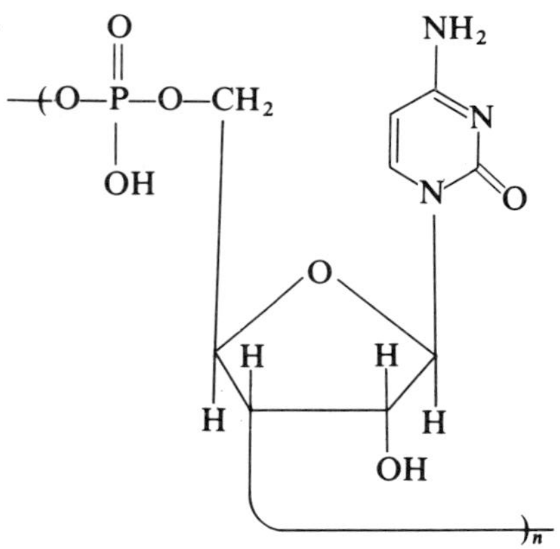	poly(ribocytidylic acid)	nucleic acids; polyesters of phosphoric acid and/or polyfurandiyls	(O, P, C)-chain	heterochain
23.	$-(\text{Ag—NC})_n-$	*catena*-poly[silver-μ-(cyano-*N*:*C*)]	—	(N, C, Ag)-chain	heterochain

9: Use of Abbreviations for Names of Polymeric Substances (1986)

The IUPAC Commission on Macromolecular Nomenclature believes that the time is opportune for a new statement regarding the use of abbreviations for the names of polymers and polymeric materials. Previous recommendations by this Commission were published in 1974 [1], and have been incorporated into an expanded draft international standard by the International Organization for Standardization (ISO) [2].

The Commission wishes to state that it has no intention, at present, of devising a new system of abbreviations, owing to the inherent difficulty of assigning systematic and unique abbreviations to polymeric structures. The Commission reminds the reader of the IUPAC policy on the use of abbreviations in the chemical literature [3], which declares, in part, that: 'there are great advantages in defining all abbreviations,..., in a single conspicuous place in each paper. This is preferably done near the beginning of the paper in a single list'.

Indeed, we urge that each abbreviation be fully defined the first time it appears in the text and that no abbreviation be used in titles of publications.

Those wishing to use abbreviations not contained in the 1974 document [1] may refer to the current ISO list [2]. Both of these lists are reproduced below, for the convenience of the reader. It is to be noted that the ISO list uses nomenclature that is not necessarily in accord with IUPAC recommendations.

LIST OF ABBREVIATIONS FROM THE IUPAC RECOMMENDATIONS 1974

PAN	polyacrylonitrile
PCTFE	poly(chlorotrifluoroethylene)
PEO	poly(ethylene oxide)
PETP	poly(ethylene terephthalate)
PE	polyethylene
PIB	polyisobutylene
PMMA	poly(methyl methacrylate)
POM	poly(oxymethylene); polyformaldehyde
PP	polypropene
PS	polystyrene
PTFE	poly(tetrafluoroethylene)
PVAC	poly(vinyl acetate)
PVAL	poly(vinyl alcohol)
PVC	poly(vinyl chloride)
PVDC	poly(vinylidene dichloride)
PVDF	poly(vinylidene difluoride)
PVF	poly(vinyl fluoride)

Reprinted from *Pure Appl. Chem.* **59**, 691–693 (1987).

LIST OF ABBREVIATIONS FROM THE 1984 DRAFT INTERNATIONAL STANDARD ISO 1043–1986*

Homopolymers and natural polymers

CA	cellulose acetate
CAB	cellulose acetate butyrate
CAP	cellulose acetate propionate
CF	cresol-formaldehyde
CMC	carboxymethylcellulose
CN	cellulose nitrate
CP	cellulose propionate
EC	ethylcellulose
EP	epoxide; epoxy
MF	melamine-formaldehyde
PA	polyamide
PB	poly(1-butene)
PBA	poly(butyl acrylate)
PBT	poly(butylene terephthalate)
PC	polycarbonate
PDAP	poly(diallyl phthalate)
PF	phenol-formaldehyde
PUR	polyurethane
PVB	poly(vinyl butyral)
PVFM	poly(vinyl formal)
PVK	polyvinylcarbazole
PVP	polyvinylpyrrolidinone
SI	silicone
UF	urea-formaldehyde
UP	unsaturated polyester

Copolymeric materials

ABS	acrylonitrile/butadiene/styrene
A/MMA	acrylonitrile/methyl methacrylate
ASA	acrylonitrile/styrene/acrylate
A/EPDM/S	acrylonitrile/ethylene-propene-diene/styrene
E/EA	ethylene/ethyl acrylate
E/P	ethylene/propene
EPDM	ethylene/propene/diene
E/VAC	ethylene/vinyl acetate
FEP	perfluoro(ethylene/propene); tetrafluoroethylene/hexafluoropropene
MPF	melamine/phenol-formaldehyde
S/B	styrene/butadiene
S/MS	styrene/α-methylstyrene
VC/E	vinyl chloride/ethylene
VC/E/MA	vinyl chloride/ethylene/methyl acrylate
VC/E/VAC	vinyl chloride/ethylene/vinyl acetate

*Partial, edited list from Reference [2].

VC/MA	vinyl chloride/methyl acrylate
VC/MMA	vinyl chloride/methyl methacrylate
VC/VAC	vinyl chloride/vinyl acetate
VC/VDC	vinyl chloride/vinylidene chloride

REFERENCES

1 IUPAC. List of standard abbreviations (symbols) for synthetic polymers and polymer materials 1974. *Pure Appl. Chem.* **40**, 473–476 (1974).
2 ISO. International Standard ISO 1043 – 1986. *Plastics – Symbols and Codes* – Part 1: Symbols for basic polymers and their modifications, and for plasticizers.
3 IUPAC. Use of abbreviations in the chemical literature (recommendations 1979). *Pure Appl. Chem.* **52**, 2229–2232 (1980).

Appendix

Bibliography of Biopolymer-Related IUPAC–IUB Nomenclature Recommendations

A AMINO ACIDS AND PEPTIDES

1 Abbreviations and symbols for the description of the conformation of polypeptide chains (tentative rules 1969). *Arch. Biochem. Biophys.* **145**, 405–421 (1971); *Biochem. J.* **121**, 577–585 (1971); *Biochemistry* **9**, 3471–3479 (1970); *Biochim. Biophys. Acta* **229**, 1–17 (1971); *Eur. J. Biochem.* **17**, 193–201 (1970); *J. Biol. Chem.* **245**, 6489–6497 (1970); *Mol. Biol.* (*Moscow*) **7**, 289–303 (1973) (in Russian); *Pure Appl. Chem.* **40**, 291–308 (1974); International Union of Biochemistry, *Biochemical Nomenclature and Related Documents*, The Biochemical Society, London, 1978, pp. 94–102.

2 Abbreviated nomenclature of synthetic polypeptides (polymerized amino acids) (recommendations 1971), *Arch. Biochem. Biophys.* **151**, 597–602 (1972); *Biochem. J.* **127**, 753–756 (1972); *Biochemistry* **11**, 942–944 (1972); *Biochim. Biophys. Acta* **278**, 211–217 (1972); *Eur. J. Biochem.* **26**, 301–304 (1972); *J. Biol. Chem.* **247**, 323–325 (1972); *Mol. Biol.* (*Moscow*) **5**, 492–496 (1971) (in Russian); *Pure Appl. Chem.* **33**, 437–444 (1973); International Union of Biochemistry, *Biochemical Nomenclature and Related Documents*, The Biochemical Society, London, 1978, pp. 88–90.

3 Nomenclature and symbolism for amino acids and peptides (recommendations 1983), *Biochem. J.* **219**, 345–373 (1984); *Eur. J. Biochem.* **138**, 9–37 (1984); *Int. J. Pept. Protein Res.* **24**, following p. 84 (1984); *J. Biol. Chem.* **260**, 14–42 (1985); *Pure Appl. Chem.* **56**, 595–624 (1984); *Spec. Period. Rep.: Amino-Acids, Pept., Proteins* **16**, 387–410 (1985). Corrections: *Eur. J. Biochem.* **152**, 1 (1985).

4 Nomenclature of glycoproteins, glycopeptides and peptidoglycans (recommendations 1985), *Eur. J. Biochem.* **159**, 1–6 (1986), corrected **185**, 485 (1989); *Glycoconjugate J.* **3**, 123–134 (1986); *J. Biol. Chem.* **262**, 13–18 (1987); *Pure Appl. Chem.* **60**, 1389–1394 (1988). (Modifies document No. 8 below by giving the short form for symbolizing oligosaccharide chains.)

B NUCLEIC ACIDS AND POLYNUCLEOTIDES

5 Abbreviations and symbols for nucleic acids, polynucleotides and their constituents (recommendations 1970), *Arch. Biochem. Biophys.* **145**, 425–436 (1971); *Biochem. J.* **120**, 449–454 (1970); *Biochemistry* **9**, 4022–4027 (1970); *Biochim. Biophys. Acta* **247**, 1–12 (1971); *Eur. J. Biochem.* **15**, 203–208 (1970), corrected **18**, 558 (1971), **25**, 1 (1972); *Hoppe-Seyler's Z. Physiol. Chem.* **351**, 1055–1063 (1970) (in German); *J. Biol. Chem.* **245**, 5171–5176 (1970), corrected **246**, 4894 (1971); *Mol. Biol.* (*Moscow*) **6**, 166–174 (1972) (in Russian); *Pure Appl. Chem.* **40**, 277–290 (1974); International Union of Biochemistry, *Biochemical Nomenclature and Related Documents*, The Biochemical Society, London, 1978, pp. 116–121.

6 Abbreviations and symbols for the description of conformations of polynucleotide chains (recommendations 1982), *Eur. J. Biochem.* **131**, 9–15 (1983); Proceedings of the 16th Jerusalem Symposium *Nucleic Acids, the Vectors of Life* (Eds. B. Pullman and J. Jortner), Reidel, Dordrecht, 1983, pp. 559–569; *Pure Appl. Chem.* **55**, 1273–1280 (1983).

7 Nomenclature for incompletely specified bases in nucleic acid sequences (recommendations 1984), *Biochem. J.* **229**, 281–286 (1985); *Eur. J. Biochem.* **150**, 1–5 (1985); *Mol. Biol. Evol.* **3**, 99–108 (1986); *Nucleic Acids Res.* **13**, 3021–3030 (1985); *Proc. Natl. Acad. Sci. U.S.A.* **83**, 4–8 (1986).

C OLIGO- AND POLYSACCHARIDES

8 Abbreviated terminology of oligosaccharide chains (recommendations 1980), *Arch. Biochem. Biophys.* **220**, 325–329 (1983); *Eur. J. Biochem.* **126**, 433–437 (1982); *J. Biol. Chem.* **257**, 3347–3351 (1982); *Pure Appl. Chem.* **54**, 1517–1522 (1982). (Document No. 4 above gives an extension of these recommendations.)

9 Polysaccharide nomenclature (recommendations 1980), *Arch. Biochem. Biophys.* **220**, 330–332 (1983); *Eur. J. Biochem.* **126**, 439–441 (1982); *J. Biol. Chem.* **257**, 3352–3354 (1982); *Pure Appl. Chem.* **54**, 1523–1526 (1982).

10 Symbols for specifying the conformation of polysaccharide chains (recommendations 1981), *Eur. J. Biochem.* **131**, 5–7 (1983); *Pure Appl. Chem.* **55**, 1269–1272 (1983).

Notes

1. Documents No. 1, 2 and 5 have been published by the International Union of Pure and Applied Chemistry (IUPAC) – International Union of Biochemistry (IUB), Commission on Biochemical Nomenclature (CBN).
2. Documents No. 3, 4, 6, 8, 9 and 10 have been published by the IUPAC–IUB Joint Commission on Biochemical Nomenclature (JCBN).
3. Document No. 7 has been published by the Nomenclature Committee of IUB.

Index